MODERN BRITISH NATURE WRITING, 1789–2020

Why do we speak so much of nature today when there is so little of it left? Prompted by this question, this study offers the first full-length exploration of modern British nature writing, from the late eighteenth century to the present. Focusing on non-fictional prose writing, the book supplies new readings of classic texts by Romantic, Victorian, Modern and contemporary authors, situating these within the context of an enduringly popular genre. Nature writing is still widely considered fundamentally celebratory or escapist, yet it is also very much in tune with the conflicts of a natural world under threat. The book's five authors connect these conflicts to the triple historical crisis of the environment, of representation, and of modern dissociated sensibility. This book offers an informed critical approach to modern British nature writing for specialist readers, as well as a valuable guide for general readers concerned by an increasingly diminished natural world.

WILL ABBERLEY is Senior Lecturer in English at the University of Sussex. His previous monographs are *English Fiction and the Evolution of Language, 1850–1914* (2015) and *Mimicry and Display in Victorian Culture* (2020). He is currently writing a book on emotions and authority in Victorian natural history literature.

CHRISTINA ALT is Lecturer in Twentieth-Century Literature at the University of St Andrews. Her research focuses on early twentieth-century exchanges between literature and science, with particular emphases on ecology, ethology and climatology. Her current research examines literary engagements with the newly formalised discipline of ecology in the modernist period.

DAVID HIGGINS is Professor of Environmental Humanities at the University of Leeds. He has published widely on Romantic-period literature and, more recently, the cultural history of environmental catastrophe. His current research investigates the philosophical genealogy of contemporary climate change discourse.

GRAHAM HUGGAN is Professor of English at the University of Leeds. His research straddles three fields: postcolonial studies, tourism studies and environmental humanities, all of which are brought together in his most recent book, *Colonialism, Culture, Whales: The Cetacean Quartet* (2018).

PIPPA MARLAND is Leverhulme Early Career Fellow at the University of Bristol. Her project 'The Pen and the Plough' explores the representation of farming in British nature writing. She has published widely on eco-poetry and creative non-fiction and is the author of *Ecocriticism and the Island* (forthcoming 2022).

MODERN BRITISH NATURE WRITING, 1789–2020

Land Lines

WILL ABBERLEY

University of Sussex

CHRISTINA ALT

University of St Andrews

DAVID HIGGINS

University of Leeds

GRAHAM HUGGAN

University of Leeds

PIPPA MARLAND

University of Bristol

CAMBRIDGE
UNIVERSITY PRESS

CAMBRIDGE
UNIVERSITY PRESS

University Printing House, Cambridge CB2 8BS, United Kingdom

One Liberty Plaza, 20th Floor, New York, NY 10006, USA

477 Williamstown Road, Port Melbourne, VIC 3207, Australia

314–321, 3rd Floor, Plot 3, Splendor Forum, Jasola District Centre, New Delhi – 110025, India

103 Penang Road, #05–06/07, Visioncrest Commercial, Singapore 238467

Cambridge University Press is part of the University of Cambridge.

It furthers the University's mission by disseminating knowledge in the pursuit of education, learning, and research at the highest international levels of excellence.

www.cambridge.org
Information on this title: www.cambridge.org/9781107191327
DOI: 10.1017/9781108123396

First published 2022

A catalogue record for this publication is available from the British Library.

ISBN 978-1-107-19132-7 Hardback

Contents

Figures

Acknowledgements

Modern British Nature Writing, 1789–2020 is a co-written book rather than a collection of individual essays. It reflects discussions over several years between its five authors, all of whom have contributed to every part of the book. Nonetheless, individuals led on particular chapters in line with their areas of specialism. David Higgins led on Chapter 1, Will Abberley on Chapter 2 and Graham Huggan on Chapter 4. Christina Alt laid the groundwork for Chapter 3, with additional contributions from Graham Huggan and Pippa Marland. Parts of Chapter 4 have been published before, in modified form, in the following journal articles: 'Affective Animals: Transspecies Encounters in Modern British Animal Writing', *Humanimalia*, 12 (2020), 210–35; 'Back to the Future: The "New Nature Writing", Ecological Boredom, and the Recall of the Wild', *Prose Studies*, 38 (2016), 152–71. Thanks are owed to the publishers of these journals for allowing us to reproduce material here. We would like to thank our respective institutions – the Universities of Bristol, Leeds, St Andrews and Sussex – for the support they have given us during the course of this project. We would also like to thank the Arts and Humanities Research Council, which sponsored the original 'Land Lines' project from which this book sprang (2017–19), as well as the two smaller projects that followed on from it (2020–1). Mike Collins deserves particular thanks for supporting and developing the project's public engagement activities. Thanks are also due to the Leverhulme Trust, which provided additional support via a Philip Leverhulme Prize awarded to Will Abberley. Finally, thanks to the various non-academic partners with whom we have had the privilege of working in connection with these projects. These include but are by no means restricted to Natural England and the National Trust, the Booth Museum (in Brighton) and the First Chances Academy (in St Andrews), and (all in Yorkshire) Yorkshire Wildlife Trust, Hatfield Woodhouse Primary School and Castle Howard Estate. This book, which began life in 2017, was eventually completed under the lockdown conditions of 2020 and early 2021. We are very grateful to Cambridge University Press for its flexibility and understanding. As is made explicit later in the book but should be evident throughout, Covid-19 has brought home to us that viruses as well as humans are part of nature, even as the pandemic has restricted our physical access to the natural world. Hard lessons remain to be learned about human-animal interdependencies as well as the proper management of our environment, and, perhaps unsurprisingly, there has been a boom in nature writing as our collective awareness of these co-dependencies has spread. British nature writing today is as popular as it has ever been, lending hope for a better future while reminding us of persistent inequalities within our society – and within a deeply divided and ecologically impoverished world. This book is dedicated to all those who seek and have sought in the past to protect and promote the natural world: all of it, and for all our sakes.

Introduction

14 July 1789 goes down in history as the day of the storming of the Bastille, the dramatic fall of which would mark the beginning of the French Revolution and the end of the *ancien régime*. Meanwhile, on the other side of the Channel in rural Hampshire, the genial parson-naturalist Gilbert White was busy tending to his orchards. 'Benham skims the horse-fields', he cheerily notes in his journal for 14 July. 'Rasps come in: not well flavoured. On this day a woman brought me two eggs of a fern-owl or eve-jarr, which she found on the verge of the hanger to the left of the hermitage, under a beechen shrubb.'[1] Robert McCrum observes that there could hardly be a greater contrast between what he questionably calls 'the resilient stability of English country life', as immortalised in White's classic study *The Natural History of Selborne* (first published in that revolutionary year), and 'the bloody metropolitan dramas of France'.[2] And yet White was also a revolutionary in his own way, a 'pioneer ecologist' whose work would lay lasting foundations for the empirical study of the natural world.[3] (A few decades later, McCrum notes, Charles Darwin would grow up with White's *Antiquities of Selborne* by his side.)

How White and the *Natural History* are seen has come in turn to shape perceptions of British nature writing. Was White an unassuming chronicler of English country ways, or was he, as his biographer Richard Mabey describes him, a complex modern figure, struggling to 'reconcile [his] love of nature with an enjoyment of the stimulation of urban life'?[4] To what extent was his sense of the stability of the natural world contradicted by his

[1] Gilbert White, 'July 14th, 1789', *The Natural History of Selborne: Journals of Gilbert White*, naturalhistoryofselborne.com/1789/07/14 (accessed 10 October 2020).

[2] Robert McCrum, '100 Best Nonfiction Books: No 80 – *The Natural History and Antiquities of Selborne* by Gilbert White', *Guardian*, 14 August 2017, www.theguardian.com/books/2017/aug/14/100-greatest-non-fiction-books-all-time-natural-history-and-antiquities-of-selborne-gilbert-white (accessed 12 December 2020).

[3] *Ibid.* [4] Richard Mabey, *Gilbert White* (London: Ebury Press, 1986), p. 3.

awareness of its capacity for change, both through human intervention and planetary forces, such as the Laki volcanic eruption of 1783? How did his nature writings reflect, whether consciously or not, the politics of landownership and usage in late eighteenth-century Britain? Is the *Natural History* no more than a well-written example of an occasional literary form – the English country diary – or is it an undisputed classic of English literature? And why does it continue to be valued, long after the countryside of which its author writes has irrevocably altered? Is reading White and similar texts from the perspective of modernity in 'one of the most nature depleted countries in the world' merely an exercise in escapism?[5]

This last question is a particularly important one for this study. Despite the remarkable efforts of many organisations and individuals, British nature is in a parlous state.[6] And yet nature writing flourishes as never before; it seems to speak to something important within the national psyche. Richard Mabey describes *The Natural History of Selborne* as 'part of that curious concoction of ideas and artefacts which are seen as somehow defining "the English way of life"'.[7] Another semi-mythical English figure, Sir David Attenborough, writes that 'the British are famous – perhaps even notorious – for their devotion to wildlife'.[8] He continues as follows:

> [T]he passion for the natural world, which can so easily become an obsession, is still widespread through British society. It leads the richest and the poorest, the humblest and the noblest, to stand for hours up to their waists in chilling salt marshes watching wildfowl, to tramp for miles across bleak moorlands just to glimpse a rare flower in bloom, to spend night after night counting birds as they fly across the face of the moon.[9]

Notwithstanding Attenborough's conflation of Englishness and Britishness – some distinctions between which are discussed later in this book – the connection is clear to a proudly amateur natural-historical tradition that has come over time to be associated, not just with the national sensibility, but with a combination of the enthusiasm and stoicism that are themselves considered features of English country life. There is a contradiction here. Indeed, the British 'devotion to wildlife' may be

[5] www.wwf.org.uk/future-of-UK-nature (accessed 23 October 2020).
[6] nbn.org.uk/wp-content/uploads/2019/09/State-of-Nature-2019-UK-full-report.pdf (accessed 23 October 2020).
[7] Mabey, *Gilbert White*, p. 6.
[8] David Attenborough, quoted in Christopher Parsons, *True to Nature* (Sparkford: Patrick Stephens Ltd, 1982), p. 7.
[9] *Ibid.*

a relatively recent invention, for it exists alongside a dark history of massive and deliberate wildlife extermination.[10] Times have changed, but valuable ecosystems are still often seen as acceptable collateral damage in the pursuit of modernisation: consider, for example, the recent destruction of ancient woodlands for the development of the HS2 railway.[11]

Nature writers and ecocritics tend to see this apparent indifference or even hostility to nature as a kind of false consciousness that might be cured through reading: a view perhaps best represented in the influential work of Jonathan Bate.[12] Here, British nature writing acts to reconnect the reader to an increasingly disenchanted world, subject both to the ravages of industrialised society and to the alienating features of modern life. Writing of this kind takes its readers on a journey of discovery, one in which their sensory perceptions of and reactions to their immediate natural surroundings are enhanced. Writing of this kind is, as Bate might put it, *poetical* rather than *political* insofar as it expresses and itself encourages a lyrical apprehension of mostly familiar natural phenomena which produces both pleasure and wonder: pleasure and wonder that are merely increased with greater scientific understanding, turning nature writing into a cross between an aesthetic primer, a biological study and a moral/spiritual guide.

This book offers a different view, based in large part on its five authors' differentiated perceptions of what marks nature writing, and British nature writing more specifically, as a *modern* form. To call British nature writing modern isn't the same thing of course as to claim that it is modern*ist*. British nature writers are rarely stylistic innovators and still more rarely political firebrands, which is by no means the same thing as saying that such writing is stylistically uninteresting or apolitical. Rather, British nature writing is modern insofar as it grapples, self-consciously at times, with the contradictions embedded within the modern condition; far from being either egocentric or ecocentric – a false division at best[13] – it struggles, against a busy backdrop of social and environmental change, to confront

10 Roger Lovegrove, *Silent Fields: The Long Decline of a Nation's Wildlife* (Oxford: Oxford University Press, 2007).

11 Julian Hoffman, *Irreplaceable: The Fight to Save Our Wild Places* (London: Penguin, 2019).

12 See, for example, Jonathan Bate, *Romantic Ecology: Wordsworth and the Environmental Tradition* (London: Routledge, 1991); also Bate, *The Song of the Earth* (Cambridge, MA: Harvard University Press, 2000).

13 The distinction is still routinely made: see, for example, Axel Goodbody, 'From Egocentrism to Ecocentrism: Nature and Morality in German Writing in the 1980s', in Catrin Gersdorf and Sylvia Mayer (eds.), *Nature in Literary and Cultural Studies: Transatlantic Conversations on Ecocriticism* (Amsterdam: Brill/Rodopi, 2006), pp. 393–415.

successive historical manifestations of the modern dissociated self. These struggles are both *ontological* – attempts to understand the nature of the self's relationship to the world – and *epistemological* – attempts to discover first principles that might then guide our knowledge of that relationship. And, tangled up with modernity as they are, they are also inevitably *representational*: they are attempts to find authentic, if not necessarily accurate, ways of representing this relationship, which is nothing if not conflicted, and all the more conflicted for being both intellectually committed and emotionally bound.

One of the premises of this book is that conflict lies at the heart of modern British nature writing, and that the anxieties it articulates are part and parcel of the modern condition itself. This conflict takes several different forms – between experience and representation, between alternative kinds of knowledge, between 'lay' and 'expert' understandings, between those who own land and those who do not – and, unsurprisingly, these forms take on specific local and historical inflections as well as feeding into general philosophical debates. It should probably go without saying that most of the key terms that underlie these debates – 'nature', 'natural world' and, for that matter, 'nature writing' – are themselves contested. 'Nature' has always been a troublesome term, meaning very different things to different people, while 'nature writers' are often suspicious of the category or reject it outright.[14] For the purposes of this book, 'nature writing' is taken as those miscellaneous forms of writing, by no means consistent over the centuries, which have consciously sought some kind of cognitive and emotional engagement with the natural world. Our focus is on non-fictional prose, sometimes seen as the defining mode of nature writing, though some attention is given to other genres and modes as well. The reason for concentrating on prose is rather different from the one provided by recent ecocritics, notably Lawrence Buell, who emphasises what he calls nature writing's 'dual accountability' – that is, to the creative imagination as well as to the various 'real-world' objects it represents.[15] This suggests a moral responsibility on the part of nature writers to say what they see and to say so as honestly as possible, but numerous filters get in the way of this, as well as the age-old problem that language effectively produces the objects it claims to represent.

[14] For a good introduction to the definitional minefield, see Kate Soper, *What Is Nature? Culture, Politics and the Non-human* (Chichester: John Wiley & Sons, 1995).

[15] Lawrence Buell, *The Environmental Imagination, Thoreau, Nature Writing, and the Formation of American Culture* (Cambridge, MA: Belknap Press, 1995), p. 91.

Similarly, Jos Smith's more recent suggestion that nature writing is a refined form of 'lyrical realism' begs several questions: not least about an apparent demand for the kind of 'fine writing' that might add stylistic lustre to standard loco-descriptive accounts.[16] As seen throughout this book, much of what potentially counts as 'nature writing' directly challenges the Romantic protocols with which it is often popularly associated, though a more accurate assertion might be that these protocols effectively challenge themselves. Much in the same way, nature writing has historically tended to engage with realist modes only to contest some of the philosophical assumptions that underlie them: it is as much an interrogation as a performance of mimesis, and as much a living testament to the non-transparency of verbal language as anything else.

Such issues are firmly embedded within non-fictional prose, especially though not exclusively the essay, the explanatory function of which is always limited by the constitutive unreliability of the language it employs. Nature writing exploits this unreliability to its own advantage: for example, by offering alternative readings of or perspectives on the same natural phenomenon or socioecological event. Nor is nature writing purely descriptive, whatever that might mean; rather it offers a nested set of *narrative* framings of natural phenomena in which the (usually) retrospective account of an eyewitness – the narrator – is folded into other possible stories and other possible readings of these stories, not least those surrounding the experiential status and interpretative credentials of the narrator himself/herself. In this last sense, nature writing probably owes more to travel writing than it lets on, though the truth value of the latter is more obviously compromised. Like travel writing, it is as much about the imagining of *journeys*, both mental and physical, as about the imagining of *place*.[17]

To what extent these part-real, part-imagined journeys offer a commentary on the *nation* remains an open question. On one hand it seems reductive to assimilate nature writing to state-of-the-nation debates or to scan it for circumstantial evidence of national allegiance; on the other it would be wrong to dissociate it entirely from what Smith calls 'preservationist nationalism' or hook it up to an evolutionary historical narrative that makes more recent nature writing seem more progressive than it is.[18]

[16] Jos Smith, *The New Nature Writing: Rethinking the Literature of Place* (London: Bloomsbury, 2017).

[17] For a more detailed discussion of this, see Patrick Holland and Graham Huggan, *Tourists with Typewriters: Critical Reflections on Contemporary Travel Writing* (Ann Arbor: University of Michigan Press, 1998).

[18] Smith, *The New Nature Writing*, p. 38.

That 'nature' operates as an indexical sign for a certain, mostly conservative idea of *Englishness*, wrapped up in comforting myths of the countryside, seems a given, but whether this sign points more broadly to *Britain* is something else. Similarly, it would be unwise to assume that nature writing simply props up these myths, which are in any case often more regionally than nationally oriented, or to speculate on how nature writing might fit or not into a national narrative that has changed, and the countryside with it, significantly over time. Perhaps because British nature writing's association with the less attractive forms of nationalism is potentially embarrassing, the topic has tended to be neglected by critics. For example, Smith's valuable study is structured around the new nature writing's complex relationships to different types of place such as the local, the wild and the archipelago, but the national is curiously absent as a category. A key problem with this absence is that it leads to a neglect of fundamentally political issues such as huge inequalities in access to nature among the British populace and the relative lack of diversity in British nature writing.[19] Questions of race, class and gender are addressed throughout this book; we focus particularly on the pervasive whiteness of British nature writing in the Afterword.

One of this book's primary tasks, therefore, is to consider what is distinctive about British nature writing and what this might say about the changing state of Britain and British cultural heritage. That nature writing is a *cultural* form seems too obvious to discuss, while it is now generally agreed that 'nature' is a *human* category – and one increasingly recognised as exhaustible in terms of the finite resources it represents. Current debates around the Anthropocene (the so-called Human Age) have raised the stakes on the protection of the natural world while also challenging what the 'natural world' might be in the first place; at their most radical, these debates have also challenged what it means to be human in what is increasingly considered a 'post-natural' world. Hence the ambivalent function of contemporary British nature writing as a multifaceted *heritage* form, shoring up traditional views of (say) the English countryside while recognising that these traditions, which are both symbolic and material in their implications, are reinvented over time. To suggest that British nature writing is a heritage form is not to

[19] For the construction of Englishness/Britishness over time, see Robert Colls, *Identity of England* (Oxford: Oxford University Press, 2002); Krishan Kumar, *The Making of English National Identity* (Cambridge: Cambridge University Press, 2003); David Matless, *Landscape and Englishness* (London: Reaktion, 1998); Mike Storry and Peter Childs (eds.), *British Cultural Identities*, 4th edition (London: Routledge, 2013); Sivamohan Valluvan, *The Clamour of Nationalism: Race and Nation in Twenty-First-Century Britain* (Manchester: Manchester University Press, 2019).

consign it to some dim-lit nostalgic realm in which 'nature', however defined, has a primarily restorative or consolatory function; rather, it is to confront the social and historical changes that are part of the fabric of heritage and the various cultural industries it supports.[20]

The charge that nature writing is inherently 'bourgeois' in sensibility and form is thus a perilous one to make (and a ridiculous one if it is applied across the board to *all* nature writing); likewise, nature writing's academic wing, ecocriticism, is well placed to fend off routine accusations that, like the cultural representations it dissects, it is either politically conservative or temperamentally inclined to pseudo-scientific pronouncements about the environment already disproved elsewhere.[21] Nature writers and/or ecocritics – so this particular argument runs – are the literary world's *faux naïfs*, suspicious of the very biological and ecological terms they draw upon, and often given to adopt an unduly reverential attitude towards what they see, with similarly misplaced faith in the referential power of language, as the 'natural world'.

It should already be clear that this book pushes against negative claims like these without necessarily seeking to recuperate nature writing as a utopian discourse, and without necessarily wanting to revive the nation-based ideologies, sometimes couched in terms of the fundamental differences between 'American' and 'British' traditions, around which it often revolves. This book seeks instead a grounded appreciation of the *imaginative* work done by nature writing in describing a realm of experience that can neither be fully captured in poetic language nor fully explained by ecological accounts of the tangled relationships between different material bodies, entities, worlds.[22] What is needed is neither an artistically enchanted nor a scientifically disenchanted but an appropriately *historicised* understanding of how nature writing operates and of how these diverse operations change over time. More specifically, this book aims at nothing less than a transformative re-examination of the history of modern British nature writing. This is not with a view to identifying, or still less reifying,

20 See, for example, Rodney Harrison, *Heritage: Critical Approaches* (London: Routledge, 2012).

21 A prominent example of this line of attack is Dana Phillips, *The Truth of Ecology: Nature, Culture, and Literature in America* (New York: Oxford University Press, 2003).

22 More is said in this book about the implications for nature writing of so-called new materialist approaches in which nature and culture are recognised as co-constitutively intertwined. 'Entanglement' and 'embodiment' are key terms in such approaches, which may share a common vocabulary but are otherwise more differentiated than is frequently supposed. A good early example, as important for its feminist approach as for its rethinking of environmental terms, is Stacy Alaimo, *Bodily Natures: Science, Environment, and the Material Self* (Bloomington: Indiana University Press, 2010).

a 'British tradition', but rather in order to interrogate some of the critical assumptions on which British nature writing, nature in Britain and Britain itself as a national 'imagined community' might be said to rest.[23]

Historicising Nature Writing

'Nature writing' is a problematic term for a study that covers more than two centuries, signifying at once a canon that authors have defined themselves in relation to (or in distinction from) and an artificial category imposed retrospectively by critics on a diverse range of texts which traverse various periods and genres. The term emerged in critical discourse at the end of the nineteenth century on both sides of the Atlantic.[24] While, for North American critics, 'nature writing' referred to the likes of Henry David Thoreau and John Burroughs, British critics applied the term first to authors such as Richard Jefferies and William Henry Hudson. In both contexts, White was invoked as a foundational influence. However, among the many things that had changed since White's time was a new conception of literature and science as opposites of each other. This conception can be gauged from Hudson's introductory remarks in *Nature in Downland* (1900), which dissociated the work from books of 'purely scientific description'. Of rural Sussex, Hudson wrote, he was concerned 'only with its smooth surface from the aesthetic point of view, and with the living garment of the downs, its animal and vegetable forms, from the point of view of the lover of nature, and, in a moderate degree, of the field naturalist'. His province, he continued, was 'impressions of the downs – of their appearance and the feelings they evoke in us'.[25]

The separation of science and the arts into discrete fields of knowledge, which Hudson felt compelled to acknowledge, had been long in the

[23] Benedict Anderson, *Imagined Communities: Reflections on the Origin and Spread of Nationalism* (New York: Verso, 2016).

[24] The earliest usage of the term 'nature writing' we have found in a British publication is in an 1894 article on Rudyard Kipling's *The Jungle Book* (1894). See '*The Jungle Book*', *Natural Science: A Monthly Review of Scientific Progress*, 5 (1894), 8–9. An 1897 letter to *The Standard* newspaper describes Jefferies as working in a tradition of 'Nature writers' descending from White: H. S. H. Waylen, 'The Work of Richard Jefferies', *The Standard* (5 February 1897), p. 4. The term seems to have taken root slightly earlier in the American context: see, for instance, 'Literature: Three Outdoor Books', *The Critic*, 24 (12 May 1894), 319–20, which dubs Henry David Thoreau, John Burroughs and others 'the "nature" writers, as they are called' (p. 319). Prior to this, Eric Lupfer notes, the American publisher Houghton, Mifflin and Company curated series of 'Out-Door Books' from the 1880s onwards: 'Before Nature Writing: Houghton, Mifflin and Company and the Invention of the Outdoor Book, 1800–1900', *Book History*, 4 (2001), 177–204.

[25] William Henry Hudson, *Nature in Downland*, 2nd edition (London: Longmans, Green and Company, 1900), pp. 20–1.

making. Through the nineteenth century, natural history societies, museum curators and, latterly, university departments had catalogued Britain's botany, zoology and geology in exhaustive detail. Travel writing as an aesthetic mode focused on personal impressions came to be opposed to the plain, factual journals of scientific expeditions.[26] Stylised, pictorial natural history illustrations gradually gave way to schematic drawings, abstract graphs and rapid photography in specialist scientific publications.[27] At the end of this period, epistemic authority over nature's facts was moving increasingly into the hands of specialist technicians working in laboratories and field stations.[28] From this perspective, White's digressive records of observations snatched in between his clerical duties looked positively quaint. As the science populariser Grant Allen commented in his introduction to an 1887 edition of *The Natural History of Selborne*, White's book was now chiefly valuable 'as a literary monument' that embalmed 'the daily life of an amateur naturalist in the days when the positions of parson, sportsman, country gentleman, and man of science were not yet incongruous'.[29] The emergence of 'nature writing' as a critical category thus coincided with the professionalisation of science and the long-term polarisation of science and literature in the anglophone imagination, which culminated in C. P. Snow's mid-twentieth-century lamentation of 'the two cultures'.[30] The label was commensurate with this trend, designating a form of *literature* concerned with subjective aesthetic and emotional experience, in contradistinction to the objective facts and methodical inferences of science.[31]

Retrospectively categorising authors of White's time and even late into the Victorian period as nature writers thus risks anachronism. Such writers worked in contexts in which science and literature intermingled. Indeed, natural history as an activity was frequently imagined as a blend of empirical observation with aesthetic and religious experience. As Lynn Merrill notes, the discourse of natural history constructed nature's objects

26 Nigel Leask, *Curiosity and the Aesthetics of Travel Writing, 1770–1840* (Oxford: Oxford University Press, 2002), p. 6.

27 Ann Shelby Blum, *Picturing Nature: American Nineteenth-Century Zoological Illustration* (Princeton, NJ: Princeton University Press, 1993), pp. 318–19.

28 See Robert E. Kohler, *Landscapes and Labscapes: Exploring the Lab-Field Border in Biology* (Chicago, IL: University of Chicago Press, 2002); Raf de Bont, *Stations in the Field: A History of Place-Based Animal Research, 1870–1930* (Chicago, IL: University of Chicago Press, 2015).

29 Grant Allen, 'Introduction', in Gilbert White, *The Natural History of Selborne* (London: John Lane, 1900), pp. xxvii–xl, xxxiv.

30 C. P. Snow, *The Two Cultures* (Cambridge: Cambridge University Press, 1998).

31 On the polarisation of the arts and the sciences into 'objective' and 'subjective' truth, see Lorraine Daston and Peter Galison, *Objectivity* (New York: Zone Books, 2007).

as 'curiosities' to be consumed and savoured much like the objects in a connoisseur's cabinet.[32] Unlike Allen's methodical 'man of science', the 'naturalist' was an expansive category that could include almost anyone with a passing interest in the natural world.[33] Models for this identity ranged from Alexander von Humboldt, who related his findings through emotionally charged impressionistic narrations, to Erasmus Darwin (grandfather of Charles), who mixed nature studies with poetic composition. Even Charles Darwin's *On the Origin of Species* (1859) remodelled nature with striking metaphors, imaginative thought experiments and a sweeping narrative of beauty and 'grandeur'.[34] For these reasons, the earlier parts of this study range across a variety of genres. Writers often mixed observations of the natural world with personal memoir and meditations and interventions on diverse topics. Writing about nature frequently served as a vehicle for exploring other issues from art and religion to politics.

That audiences recognised the hybridity of such writing about nature is shown in a reviewer's comment on Jefferies that the author 'tells us in a discursive and half-poetical but very attractive way what he has seen. He does a great deal to teach natural history without formality. Those who will study it under him will find it only a pleasure . . . To infect a young person with this sentiment is to educate in the truest sense.'[35] The designation of Jefferies' writing as 'half-poetical' reflected his rigorous commitment to facts, recording original data on animals' habits and living conditions and dispelling popular myths about them. Yet, at the same time, he presented such investigations as motivated by and stimulative of pleasure and the moral-spiritual imagination.

Lawrence Buell groups such cross-genre material in the North American context under the category of 'literary naturism', designating 'an interest in representation of literal nature as a substantial if not exclusive part of one's literary project'.[36] Yet in Britain, prior to the end of the nineteenth century, detailed literary engagements with 'literal nature' were rarely so easily distinguished from other forms. While in North America, Henry

[32] Lynn L. Merrill, *The Romance of Victorian Natural History* (Oxford: Oxford University Press, 1989), pp. iii–vi.

[33] Ruth Barton, '"Men of Science": Language, Identity and Professionalization in the Mid-Victorian Scientific Community', *History of Science*, 41 (2003), 73–119 (p. 103).

[34] See Gillian Beer, *Darwin's Plots: Evolutionary Narrative in Darwin, George Eliot and Nineteenth-Century Fiction* (Cambridge: Cambridge University Press, 2009); George Levine, *Darwin the Writer* (Oxford: Oxford University Press, 2011).

[35] '"Nature near London." By Richard Jefferies', *British Quarterly Review*, 78 (July 1883), 212.

[36] Buell, *Environmental Imagination*, p. 431.

David Thoreau did much to set the model for the nature essay, British nature writing lacked a pioneer of comparative stature.[37] As Chapter 2 shows, Jefferies' essays often echo Thoreau's detailed attention to natural environments and his depth of feeling, but they also diverge from Thoreau's wilderness: an imagined pure wild nature untainted by human society. Jefferies pays as much attention to the human inhabitants of the countryside and their customs as he does to animals and plants. Indeed, if nature writing is tied to North American experiences and ideas of the wilderness, then British writers (at least when focused upon Britain) have been supplied with materials for something else, the lands around them marked by many centuries of cultivation, enclosure and economic activity. In addition, neither Jefferies nor others following him produced a single book that defined the genre as *Walden* did. Tracing the roots of British nature writing thus requires a wide and roving lens and scepticism of Buell's facile division between the 'naturist' writer and the 'naturalist . . . belonging to the natural history tradition'.[38]

Furthermore, nature writing since the late nineteenth century has rarely sat in such neat contradistinction to science as Hudson's statement implies. Indeed, in the same text, Hudson challenges scientific orthodoxy on a number of points and sets himself up as an authority on Sussex's landscapes by virtue of his long acquaintance with them. His self-identification as a 'field naturalist' points to the openness of the field as a space of study, an openness that extended well into the twentieth century. While anatomy, morphology and physiology increasingly focused on the activities of museum closets and university laboratories, amateurs continued to be influential in field observation, particularly of wild animals.[39] At the same time, methodical study and wandering and leisurely curiosity could overlap in activities such as birdwatching, which emerged as a popular hobby around the turn of the twentieth century. The seemingly infinite nature of the field, in which one habitat merges with another, draws attention to the reductiveness of scientific efforts to cut nature up into discrete objects and quantities. The irreducibility of life, long recognised by nature writers, would be mirrored in the development of the disciplines of ecology and ethology (see Chapter 3).

Allen's description of White's parsonical ornithology as a charming anachronism for late nineteenth-century readers defined natural science in contradistinction to religion as well as to literature. Literature and

[37] Merrill, *Romance*, p. 13. [38] Buell, *Environmental Imagination*, p. 341.
[39] See Kohler, *Landscapes and Labscapes*, pp. 80–7.

religion often find themselves on the same side of such dichotomies due to their identification with the imagination, a faculty sometimes reviled as the enemy of scientific factuality and sober, sceptical rationalism.[40] However, nature writing channelled traditions that opposed such dichotomies. For example, since the eighteenth century, natural theology had sought to marry scientific knowledge with religious faith. Douglas Hedley notes that theistic science necessarily implies imagining more than we can empirically demonstrate or rationally explain. Belief in a 'more-than-material' existence, he argues, renders the world partly figurative, filling it with meaning and value beyond its physical properties and practical uses.[41] Rationalist materialism was further countered by Romantic reformulations of the imagination as a productive – rather than merely reproductive – power. Immanuel Kant's model of the universe as partly created by the mind lent the imagination an epistemic authority independent of external facts. For influential authors such as Samuel Taylor Coleridge, the transcendent power of imagination allowed for knowledge of nature to be linked with humanity's inner experiences of beauty, moral feelings and intuitions of extrasensory realities.[42] Tied to this Romantic idealism was a view of nature as a dynamic 'process of becoming' that resisted purely mechanistic description and could only be fully appreciated through the holistic interaction of all human faculties.[43]

Critics of American nature writing have linked these influences to an early ecological ethics, notably manifested in a Thoreauvian 'animism' that invests the natural world with spiritual value.[44] Such animism supposedly diverges from the Cartesian tradition of opposing matter to mind or spirit, framing the spiritual as immanent rather than transcendent. John Gatta places Thoreau's animism within a wider trajectory in the history of American culture by which reverence for the natural world increasingly

[40] See Lorraine Daston, 'Fear and Loathing of the Imagination in Science', *Daedalus*, 127 (1998), 73–95.

[41] Douglas Hedley, 'Imagination and Natural Theology', in John Hedley Brooke, Russell Re Manning and Fraser Watts (eds.), *The Oxford Handbook of Natural Theology* (Oxford: Oxford University Press, 2013), pp. 539–50 (p. 542).

[42] *Ibid.*, pp. 546–8.

[43] Kate Rigby, *Topographies of the Sacred: The Poetics of Place in European Romanticism* (Charlottesville: University of Virginia Press, 2004), pp. 20–5.

[44] Lawrence Buell, 'Introduction', in *Thoreau's Sense of Place: Essays in American Environmental Writing* (Iowa City: University of Iowa Press, 2000), pp. 1–14 (p. 6); Bron Raymond Taylor, *Dark Green Religion: Nature Spirituality and the Planetary Future* (Berkeley: University of California Press, 2010), p. 51; Michael P. Nelson, 'Animism', in Shepard Krech, J. R. McNeill and Carolyn Merchant (eds.), *Encyclopedia of World Environmental History* (New York: Routledge, 2004), pp. 55–6.

substituted for Christian belief, undermined by biblical criticism and secular science.[45] A similar tendency to sacralise wild objects and spaces might be traced through the British nature writing tradition, from William Wordsworth's 'spots of time' to John Ruskin's celebrations of nature's divine beauty and Richard Jefferies' heathen reveries on Wiltshire hills to Robert Macfarlane's solemn pilgrimages along ancient rural footpaths. Chapter 2 discusses how Victorian nature writing that argued for the conservation of wild environments and species typically appealed to a vaguely spiritual sense of value, associating ecological destruction with soulless materialism and philistinism. Religious discourse continues to haunt twenty-first-century British nature writing's valuation of the wild beyond instrumental utility. As one interview quotes Macfarlane on his emotional responses to landscapes: 'The term "spiritual" is so culturally contraband. But I've searched and searched, and there are no better words. Religious language is all we've got.'[46]

However, it is not always clear how much the spiritual language of British nature writing has realised an ecocentric ethics. Although such rhetoric does not have to imply Cartesian dualism, it has often had the effect of devaluing the material basis of life.[47] Hence, Jefferies' hillside reveries are sometimes more concerned with transcending nonhuman, physical 'nature' than sympathising with it (see Chapter 2). Even when it avoids dualism, spiritualist rhetoric in the Western tradition has frequently implied a hierarchy in which humans represent Earth's most spiritually exalted beings and, therefore, the most important.[48] It is worth asking, then, whether nature writing (British or otherwise) can ever escape the orbit of anthropocentrism. Even when nature writers argue for the value of environments and species, where exactly are they locating this value: in nonhuman life per se or in humans' experiences of it? Romantic reverence for subjective experience has long been attacked (although not always fairly) as a vehicle for narcissism, and the influence of this tradition upon British nature writing lays the form open to the same critique. In a polemic against contemporary nature writers, Richard Smyth complains: '"This

[45] John Gatta, *Making Nature Sacred: Literature, Religion, and Environment in America from the Puritans to the Present* (Oxford: Oxford University Press, 2004), pp. 127–42.

[46] Quoted in Rachel Cooke, 'Robert Macfarlane: "Paths Are Human; They Are the Traces of Our Relationships', *Observer*, 27 May 2012, www.theguardian.com/books/2012/may/27/robert-macfarlane-old-ways-interview (accessed 12 December 2020).

[47] See Lynn White, 'The Historical Roots of Our Ecological Crisis', *Science*, 155 (1967), 1203–7.

[48] See Gary Steiner, *Anthropocentrism and Its Discontents: The Moral Status of Animals in the History of Western Philosophy* (Pittsburgh, PA: University of Pittsburgh Press, 2005), pp. 53–93.

landscape is special" can shade into "I am special."'[49] Laying aside personal narcissism, it is undoubtedly true that valuing nonhuman life as a source of emotional and aesthetic stimulation tends to centre on human interests. In this way, nature writing exemplifies the inherent anthropocentrism of all literature by virtue of its reliance on human language.

Nonetheless, if worshipping nature also involves worshipping ourselves, this outcome is not necessarily inimical to ecological politics if it collapses the human/nature binary. Defending Emerson against the criticism that the author was more interested in his own thoughts than nature in itself, Nicholas Aaron Friesner argues that his ambivalent 'ideal of intimacy with the world preserves a unique sense of human responsibility while rejecting any picture in which human moral activity is autonomous from the environments it occurs'. Friesner contends that the ecocritical tendency to privilege 'empirical' outlooks over 'metaphysical spiritualist' ones reinforces Cartesian dualism, implying 'a nature that exists waiting to be uncovered' and downplaying the ways in which nature is physically and conceptually constructed by humans and entangled with social and political realities.[50] Similarly, the inclination of authors since Ruskin (who was strongly influenced by Emerson) to celebrate Britain's more-than-human life and environments as spiritually enriching national treasures has challenged the nature/culture opposition, emphasising humans' entwinement with the wider living world. At the same time, nature writers may counteract the inherent anthropocentricity of writing by deprecating their own efforts to capture the world in words.[51] Numerous authors discussed here present their efforts to write nature as necessarily doomed, or at least inadequate. Yet acknowledging language's incapacity to conquer nature can also be an opportunity to reimagine language, thought and aesthetics, as patterns within wider, non-anthropocentric networks of life.

Such ecological consciousness would frame nature writing as not only an aspect of the organic world, but also as a form of agency. This can cut both ways. The son of Philip Henry Gosse, Edmund, observed that, by eulogising the beauties of Britain's coastal wildlife, his father's books had encouraged the destruction of these same environments by crowds of reader-collectors: a bitter irony that caused his father 'great chagrin' in

49 Richard Smyth, 'The Cult of Nature Writing', *New Humanist* (Spring 2016), newhumanist.org.uk/articles/5019/the-cult-of-nature-writing (accessed 12 December 2020).

50 Nicholas Aaron Friesner, 'A Transcendentalist Nature Religion', *Religions*, 8 (2017), 1–18 (pp. 2, 6–7).

51 George Seddon, *Landprints: Reflections on Place and Landscape* (Cambridge: Cambridge University Press, 1997), pp. 16–18.

old age.[52] By representing wild environments, Gosse's popular writings had ultimately altered them. Yet nature writers would also come to model responsible behaviour in such environments. The sentimental responses of authors such as Ruskin and Hudson to Britain's wild populations would align with the efforts of early conservation movements such as the Royal Society for the Protection of Birds and the National Trust to limit anthropogenic habitat destruction and species extinction (see Chapters 2 and 3). In this sense, British nature writing seeks not only to reflect the natural world, but also to imagine alternative, possibly more desirable natural worlds that might be brought into being. As the anti-capitalist firebrand Ruskin realised, idealism is a precondition of utopianism. Nature writing continues to play an important, although sometimes controversial role in promoting changing notions of conservation: consider, for example, works such as Benedict MacDonald's *Rebirding: Restoring Britain's Wildlife* (2020), George Monbiot's *Feral: Rewilding the Land, Sea and Human Life* (2014) and Isabella Tree's *Wilding: The Return of Nature to a British Farm* (2018).

In recent years, scholars of nineteenth-century British literature have argued for a 'strategic presentism' that puts past and present in dialogue, using historical texts as points of departure to approach current environmental issues from fresh perspectives.[53] This book recognises the value of such efforts and similarly makes use of contemporary concepts to illuminate the complex, changing ways in which writers since 1789 have constructed the natural world in relation to British culture. However, it also seeks to acknowledge the alterity of the past, showing how different historical contexts have constructed different natures. Although, in some ways, White's work looked forward to later conceptions of wildlife as holistic ecosystems, it was also shaped by a tradition of natural theology with roots in constitutional monarchism and political economy.[54] In contrast to the oppression, disorder and bloodshed associated with religious factionalism, English natural theology had sought a broad,

[52] Edmund Gosse, *Father and Son: Biographical Reflections* (New York: Charles Scribner's Sons, 1907), p. 150.

[53] See David Sweeney Coombs and Danielle Coriale, 'V21 Forum on Strategic Presentism', *Victorian Studies*, 59 (2016), 87–9; Jesse Oak Taylor, *The Sky of Our Manufacture: The London Fog in British Fiction from Dickens to Woolf* (Charlottesville: University of Virginia Press, 2016); Wendy Parkins and Peter Adkins, 'Introduction: Victorian Ecology and the Anthropocene', *19: Interdisciplinary Studies in the Long Nineteenth Century*, 26 (2018), 19.bbk.ac.uk/article/id/1717/ (accessed 12 December 2020).

[54] See Nigel Cooper, 'Gilbert White and the Natural Theology of Selborne', *Linnean*, 23 (2007), 29–44.

Latitudinarian consensus based on indisputable natural facts.[55] It had also grown to endorse political liberalism, suggesting that God had arranged the world with the intention of humans cultivating it and developing commerce and industries.[56] Thomas Malthus was thus able to present his *Essay on the Principle of Population* (1798) as a reading of not only facts, but also divine values. The natural world he depicted, characterised by struggle and starvation, was held to teach hard work and self-reliance and to condemn idle pleasure-seeking. These theological and ideological associations would continue to haunt later representations of nature after Darwinian evolution by natural selection further enshrined competition as a law of life. Malthus' legacy continues to the present day in the often virulent controversies around the significance of 'overpopulation' in producing climate change and biodiversity loss.[57]

At the same time, interpretations of nature varied depending on the interests of the interpreter. Malthus' *Essay* was written to oppose arguments for endless progress, which were rooted in the natural-theological view of providence guiding societal development to maximise happiness. Similarly, socialists and anarchists would later cite evolutionary models of nature as vindications of their values of equality and cooperation.[58] Further, White's natural theology was not simply replaced by atheistic materialism. The idea that physical nature was only a kind of surface to deeper, spiritual forces persisted into twentieth-century biology through vitalist arguments for an immaterial element or power that distinguished organic matter from inorganic.[59] Beyond these vitalist–materialist disputes, inquirers have long been torn between mechanistic and organicist models of nature. While the former attitude assumes a 'clockwork' universe that is ultimately predictable and reducible to component elements and mechanisms, the latter contends that organisms and the larger systems they inhabit are greater than the sums of their parts.[60] Although organicism

[55] Robert J. Mayhew, *Landscape, Literature and English Religious Culture, 1660–1800: Samuel Johnson and Languages of Natural Description* (Basingstoke: Palgrave, 2004), pp. 70–126.

[56] Paul Oslington, *Political Economy As Natural Theology: Smith, Malthus and Their Followers* (London: Routledge, 2017).

[57] For one view, see Eileen Crist, *Towards an Ecological Civilisation* (Chicago, IL: University of Chicago Press, 2019).

[58] See Eric. M. Johnson, 'The Struggle for Coexistence: Peter Kropotkin and the Social Ecology of Science in Russia, Europe, and England, 1859–1922', PhD thesis (University of British Columbia, 2019).

[59] Geert Jan M. de Klerk, 'Mechanism and Vitalism. A History of the Controversy', *Acta Biotheoretica*, 28 (1979), 1–10.

[60] Erik L. Peterson, *The Life Organic: The Theoretical Biology Club and the Roots of Epigenetics* (Pittsburgh, PA: University of Pittsburgh Press, 2017).

originated in Romantic philosophy, it has come to underpin ecological conceptions of nature as a holistic whole of interrelated processes. Natural theology and the natural history that grew out of it could be strongly mechanistic, as William Paley demonstrated in his influential 1802 image of the world as a finely tuned watch.[61] However, their emphasis on life forms' endless interrelations through shared environments could also promote organicist thinking.

Contemporary theorisations of nature writing as an inherently ecological genre thus point to the form's partial roots in natural history, which focuses on tangible objects and perceptible phenomena in contrast to the more abstract, mathematical tendencies of other natural sciences. By foregrounding the experiences and perspectives of observers, natural history can remind us of our embeddedness within the nature we attempt to describe. Science no longer resembles a window upon the natural world from some objective 'outside', emerging instead as an activity by some of that world's inhabitants, shaped and circumscribed by environments. The tradition of natural history can also resist reductive mechanicism through its historical consciousness. Devin Griffiths notes that British natural history grew out of a discourse of 'comparative historicism', which, by juxtaposing different epochs, accentuated the roles of chance in history.[62] The imaginative character of natural history, conjuring vanished pasts from clues in the landscape, highlights not only the temporary nature of the environmental present, but also its contingency. The natural world we see around us could easily have been different, and its future is undetermined. In practice, nature writing is not always ecologically convincing, perhaps particularly when at its most self-involved. But the best nature writers are often accomplished amateur naturalists and some have considerable scientific training.

Nature Writing As Labour and Leisure

Two obvious generic (or modal) influences upon British nature writing are the pastoral and the georgic.[63] While the pastoral tradition tends to construct rural environments as spaces of leisurely ease and contemplation,

[61] William Paley, *Natural Theology: or, Evidences of the Existence and Attributes of the Deity* (London: J. Faulder, 1809), pp. 1–2.

[62] Devin Griffiths, *The Age of Analogy: Science and Literature between the Darwins* (Baltimore, MD: Johns Hopkins University Press, 2016), pp. 2–6, 9–10.

[63] On the genre/mode distinction, see Paul Alpers, *What Is Pastoral?* (Chicago, IL: University of Chicago Press, 1996), p. 50.

the georgic focuses on the physical labour by which humans adapt the earth to their wants. David Fairer notes that these different emphases shape different ways of depicting the countryside. While the pastoral is associated with 'ideal', non-specific, unchanging landscapes and spiritual abstractions, the georgic gravitates towards temporal and spatial particularity, the hard facts of life, changing of seasons and artificial modification of landscapes. For Fairer, the georgic is distinctively 'capacious', ranging across various times, places and themes. This capaciousness renders the georgic 'an appropriate mode for expressing the energies of trade and colonization' in contrast to the pastoral's more feudal, aristocratic associations.[64] Any hard distinction between the two genres, however, tends to elide the complex historical realities that informed them, such as the hugely disruptive effects of enclosure on British rural life in the eighteenth and nineteenth centuries (see Chapter 1). Georgic representations, such as in Oliver Goldsmith's influential 'The Deserted Village' (1770), could be just as idealised as pastoral ones.

These strains of influence have shaped persistent ambivalences through the history of British nature writing. What exactly is the nature writer seeking or celebrating? Harmony with nature or domination over it? At times, British nature writers construct the object of their inquiries as an authentic, timeless wilderness unspoilt by human meddling and, at others, as an elaborate garden bearing generations of human management and adaptation. Consider, for example, Ruskin's reverence for wild birds in contrast to Jefferies' aestheticization of the threshing machine (see Chapter 2). Pastoral and georgic elements run through British nature writing to the present day, not least because the most interesting writing in both modes has always contended with the threat of loss and the fear that human-nonhuman harmony can only be found in the past. Much British nature writing draws on this tendency to elegiac retrospection, while the georgic's capaciousness is echoed in the 'wandering' narrative form that is often adopted. Further uncertainty obtains regarding the place (or displacement) of the nature writer/narrator in environments designated as 'natural'. The nature writer can be at once an idler consuming the landscape like a fine wine and a worker producing it through disciplined

64 David Fairer, 'Persistence, Adaptations and Transformations in Pastoral and Georgic Poetry', in John Richetti (ed.), *The Cambridge History of English Literature, 1660–1780* (Cambridge: Cambridge University Press, 2008), pp. 259–86. On the potential opposing class associations of georgic and pastoral, see Anthony Low, *The Georgic Revolution* (Princeton, NJ: Princeton University Press, 1985); Kevis Goodman, *Georgic Modernity and British Romanticism: Poetry and the Mediation of History* (Cambridge: Cambridge University Press, 2005).

effort, merging with the amorphous identity of the 'naturalist'. Jefferies and Hudson often resemble the first type, sampling the delights of the landscape while the rural working class toils in it, oblivious to such pleasures. Conversely Hugh Miller presents his appreciation of Scotland's rocky scenes as a product of his labouring amongst them as a quarryman and stonemason (see Chapter 2) while John Stewart Collis argues that agricultural work opens his eyes to fresh perspectives on the land (see Chapter 3).

These equivocations between leisure and labour in nature writers' self-presentations index how economic and technological developments shaped different social classes' relationships with the countryside and nonhuman life. The early period covered by this study witnessed an intensification of agricultural enclosures that – whatever their overall benefits – enriched elites while severely attenuating the common rights of rural labourers. A little later, the emergence of a national railway network enabled fast, relatively cheap access to the countryside even as the population balance shifted increasingly towards towns and cities. At the same time, steadily growing literacy, disposable incomes and printing technologies opened up new readerships and opportunities for professional writers. Nature writing participated in this popularisation of print culture, above all in the many editions printed of White's *Natural History* that made it reputedly the fourth most published book in the English language.[65] Nature writing offered at once vicarious experiences of rural environments and guides to these environments. Further, authors such as Miller, Gosse and Hudson rendered nature more accessible by finding pieces of it interpenetrated with urban environments. As Chapter 2 demonstrates, such writers presented nature studies as a form of 'rational recreation' which blurred the boundaries between work and play. Yet nature writers could also set themselves and their implied readers apart from the masses, presenting themselves as cultivated aesthetes in contradistinction to the vulgar. Authors could advocate the study and appreciation of nature as open to all while simultaneously modelling these activities as markers of social distinction. Tensions between an idealised vision of nature for all and the class-based realities of unequal access and usage continue to resonate in the present day and have been brought into sharp relief by the Covid-19 pandemic.

[65] Mark Daniel, 'Introduction', in *The Essential Gilbert White of Selborne* (London: Breslich & Foss, 1983), p. x.

The period covered by this book has seen huge changes to the British landscape and land use. And yet landownership itself remains surprisingly feudal. As Guy Shrubshole has shown, half of England is owned by less than 1 per cent of the population, with aristocrats and gentry providing the largest bloc (30 per cent) of a relatively small dominant group of 25,000 landowners.[66] This matters for all kinds of reasons. The history of nature writing is – or should be – also a history of inequality between social classes, between genders, between ethnicities and between humans and nonhumans. If we consider British nature writing as in part a heritage form, it is worth recalling (from nearly fifty years ago) Raymond Williams' remarks on the fashion for visiting country houses:

> Stand at any point and look at that land. Look at what those fields, those streams, those woods even today produce. Think it through as labour and see how long and systematic the exploitation and seizure must have been to rear that many houses, on that scale . . . What these 'great' houses do is to break the scale, by an act of will corresponding to their real and systematic exploitation of others . . . Much of the real profit of a more modern agriculture went not into productive investment but into that explicit social declaration: a mutually competitive but still uniform exposition, at every turn, of an established and commanding class power.[67]

We can extrapolate two points from this passage that are of particular relevance to this book. First, not only country houses, but also rural landscapes bear the marks of dispossession and inequality. One might think, for example, of the ecologically denuded state of many of Britain's national parks and the huge areas taken up by intensively managed grouse moors. These moors may benefit a few upland species and provide some rural employment, but overall their ecological and probably economic effects are well into the negative. Their management often involves the persecution of predator species, as well as the burning of heather and peat, which can increase flood risks in the lowlands, increase carbon emissions and damage precious habitats. Like the inhabitants of Williams' country houses, those who benefit from driven grouse shooting are not those who bear the burden of its considerable costs.

The second extrapolation is more complex. The 'more modern agriculture' created by enclosure certainly did not benefit everyone equally. It led

[66] Guy Shrubshole, *Who Owns England?* (London: HarperCollins, 2019). Recent data for other parts of the United Kingdom are harder to come by, although see Andy Wightman, *Who Owns Scotland* (Edinburgh: Canongate, 1996).

[67] Raymond Williams, *The Country and the City* (London: Vintage, 2016 [1973]), p. 151.

to large numbers of effectively dispossessed rural people moving to urban areas where they provided cheap labour for industrialisation. At the same time, Britain's industrial power was also fuelled by resources and wealth extracted from its colonies. The dispossessions experienced by the British peasantry and those experienced by indigenous peoples in colonised lands were not equivalent, but were governed by a similar logic of extractivist 'improvement' that tended to benefit elites. Furthermore, these internal and external dispossessions were mutually reinforcing by allowing power and wealth to consolidate. The history of the British landscape, therefore, cannot be disconnected from the history of British imperialism, or for that matter the nation's contribution to environmental change at a global level. Due to its emergence as the first global industrial power, up until the early twentieth century, the United Kingdom had accumulated more CO_2 emissions than any other country in the world. Even now it is the fifth largest emitter in historical terms.[68]

British nature writing has at times, not unjustly, been accused of not being sufficiently political.[69] But there are different ways of doing politics. One might see its close attention to the nonhuman world and its frequent attempts to push against self-interested anthropocentrism as themselves political acts. Its debt to the pastoral and georgic can be seen in its attempts to articulate ecological ethics. Bridget Keegan argues that the georgic's sense of humanity's reliance on the land inculcates an ethic of stewardship in contrast to the pastoral's consuming gaze.[70] Similarly, Donald Worster regards eighteenth-century notions of nature as an integrated 'economy' (linked with the revival of georgic poetry in the period) as embryonic of ecological thought.[71] At times, writers such as Jefferies can seem to demonstrate this ecological georgic as they depict human and nonhuman lives as imbricated and mutually dependent. However, while pastoral idealisations have obvious dangers, the influence of the georgic can also be felt in distinctively unecological passages, such as Charles Kingsley's celebration of the draining of marshlands and eradication of their beaver populations (see Chapter 2). In such moments, the georgic and its focus on labour seem wedded to a utilitarian, extractivist vision of the organic world. The georgic

68 For a useful visualisation produced by Carbon Brief, see www.youtube.com/watch?v=jx85qK1ztAc (accessed 19 November 2020).

69 See, for example, Mark Cocker, 'Death of the Naturalist', *New Statesman*, 17 June 2015.

70 Bridget Keegan, *British Labouring-Class Nature Poetry, 1730–1837* (Basingstoke: Palgrave, 2008), p. 14.

71 Donald Worster, *Nature's Economy: A History of Ecological Ideas* (Cambridge: Cambridge University Press, 1977).

may therefore seem to an apt form for the Anthropocene. It takes human-nonhuman entanglements as a given, but offers very different versions of the associated power relations, ranging from dominance to stewardship to cohabitation. Given that more than 70 per cent of the United Kingdom's land is devoted to agriculture, and that this figure is unlikely to decrease any time soon, understanding how farmed land can better make space for nonhuman nature is vital if we are going to rectify the huge decline of British wildlife populations.[72] British nature writing has sometimes unhelpfully framed farming and farmers as the enemy, although there are signs that this is changing with the emergence of authors such as James Rebanks and Isabella Tree.

Method and Structure

This book is absolutely not intended as a survey. We take a selective approach so as to examine how British nature writers since White have engaged with modernity and its conflicts. While we have done our best to ensure that our discussion is appropriately wide-ranging and our selected texts appropriately diverse, we have taken as read that detailed textual analysis should drive this study. In order to bring out the complexity and significance of British nature writing, we have therefore aimed for depth rather than range. Our focus is largely on prose. As previously discussed, the prose essay is particularly suited to nature writing due to its precarious balancing act between explanation and imagination; we also note that the popularity and influence of Anglo-American prose nature writing in the present day far outweighs that of poetry. Nonetheless, we have not arbitrarily banished poetry from our discussion. Some of our best prose nature writers from Charlotte Smith to Kathleen Jamie are also accomplished poets, and some of our key texts move creatively between the two modes. Readers will no doubt think of authors who do not feature, but whom we might have included. This is inevitable given this book's historical coverage and the diversity and amorphousness of nature writing as a form. We firmly believe, however, that our approach has been the right one and are confident that readers will find new things in this study: new ways of understanding the value and power of nature writing, its engagement with

[72] www.theguardian.com/environment/2019/dec/31/convert-farmland-to-nature-climate-crisis (accessed 29 October 2020). Improved farming practices of course can also have a range of other environmental benefits such as carbon sequestration.

modernity, and the complexity and importance of familiar and unfamiliar writers.

Like our primary texts, our methodological and theoretical influences are wide-ranging. Within English studies, we draw particularly on two key areas. From ecocriticism, we have taken an understanding of literary language not merely as representing the world, but as entangled with it. We are particularly responsive to so-called material ecocriticism with its emphasis on nonhuman agency and 'storiedness'. And from literature and science we understand the two cultures not as separate but as porous, interconnected and often sharing tropes and narrative structures. We also draw on broader interdisciplinary fields. From cultural geography, for example, we take a concern with how landscapes are simultaneously real and imagined spaces. Much of our analysis concerns attempts to see the world from nonhuman perspectives and to push against the split between nonhuman nature and human culture. We therefore draw on work in animal studies that challenges anthropocentric framings of the human-nonhuman relationship. Finally, drawing on the environmental humanities, we understand questions that to past scholars might have seemed entirely human ones – around value, justice, epistemology and so on – as always ecologically embedded.

We take a chronological approach to British nature writing, with our four chapters divided into loosely defined periods: Romantic, Victorian, Modern and Contemporary. Authors' lives and works do not always fit neatly into these kinds of chronological segments, of course, and we are also concerned throughout with identifying continuities across periods. Thus some authors span chapters and figures analysed in detail in the earlier part of this book reappear due to their influence on subsequent writers. Such overlaps reveal common thematic concerns as well as shedding light on nested sets of relationships: between human and nonhuman, nature and culture, experiences and representations of the natural world. As might be expected in a book of this scope, the insights it provides cannot be reduced to a single line of argument, and as might be expected from a form as diverse and polymorphous as nature writing, the work we do produces no definitive understanding or standard definition of nature writing as a literary genre. For all that, this book is driven, at least in part, by a view of nature writing that pushes against popular conceptions of the genre as consolatory: as a vehicle for pleasure and wonder that soothes our anxieties (whoever 'we' may be at any given moment) about the stresses and challenges of the modern world. On the contrary, as we have already begun to argue, nature writing *articulates* these anxieties. It is a conflict-ridden mode that speaks to the contradictions embedded within the modern condition and a crisis-ridden mode that addresses, whether directly or indirectly, some of the crises that have come over time to be associated with modernity: the crisis of the environment, the crisis of representation, the crisis of the alienated self.

This book is concerned, in other words, to dispel once and for all the idea that nature writing is essentially a compensatory form with only limited relevance to the times we live in. It is no longer possible – if it ever was – to see nature writing as a 'niche' form, or as one unworthy of serious academic study. And it is no longer possible – if it ever was – to see nature writing as an 'escapist' form; rather, it engages, never more so than in the present, with issues of the utmost importance to the future life of the planet – and to our own future lives. Consequently, this book is wedded to the view that nature writing can and does make a difference, not least by offering us new ways of imagining the world to which literature and the other arts are central, and by creating new stories about our changing relationship with the environment – which are also stories we tell about ourselves. Nature writing has always had a consciousness-raising role, although by no means all of the works we look at in the following pages can or indeed should be classified in present-day terms as 'environmentalist'. As will become quickly apparent, the genre continues to this day to struggle with basic issues of equality and inclusion that are products of specific (national) social hierarchies as well as of an unevenly developed world. Today, at a time of accelerated climate change and mass species extinctions – at a time in which, most recently, the coronavirus outbreak has revealed to us once again the interdependence of animal and human welfare for each other's survival – this role is probably as important as it has ever been. In this contemporary context of emergency – one which, as we will see, is repeatedly foreshadowed in the work of earlier writers – nature writing performs *both* an aesthetic and a political function: for example, in encouraging us to seek social as well as ecological justice, and in urging us to imagine more equitable and sustainable ways of living in a coinhabited world that for far too long we have been fashioning after far too few of ourselves.

Chapter 1 debunks lazy clichés about Romantic nature writing by revealing its diversity, complexity and self-consciousness. It brings together canonical and non-canonical writers who share a concern with 'nature in motion' and therefore have something in common with present-day new materialist thinkers. Like the rest of this book, its principal focus is on non-fictional prose; however, it also addresses the porousness of nature writing in the period, when it is not so much an identifiable genre as a mode of engagement across and within different kinds of texts, including those that combine poetry and prose. As one would expect, we start with Gilbert White, but complicate British nature writing's origin story by examining him alongside Jean-Jacques Rousseau. Our claim is that, while White's localism and close field observations were certainly influential on later nature writers, so was the strand of confessional autobiography pioneered by Rousseau. Romantic nature writing is often represented as primarily a self-aggrandising masculine mode. We challenge this narrative by showing how important women writers were to its development. We focus particularly on Dorothy Wordsworth and Charlotte Smith, both of whom experimented with literary form in an attempt to represent the dynamism of nonhuman nature: not as wild and separate, but as entangled with everyday human life.

It is unsurprising that John Clare would feature heavily in a chapter on Romantic nature writing, given his reputation as a peerless literary observer of creatures and natural phenomena. However, we shed new light on the significance of his natural history prose, which has tended to be neglected in favour of his poetry. We also bring Clare into dialogue with two other rural labouring-class figures whose roles in the history of British nature writing have yet to be fully acknowledged: the natural historian and wood engraver Thomas Bewick and the novelist and poet James Hogg. All three authors resisted and lamented the forces of modernisation, but did so through developing innovative modes of representation that appealed to a self-consciously contemporary urban readership. Even at its most backward-looking, Romantic nature writing engaged with the ontological, epistemological and representational contradictions that we explore throughout this study. We conclude with some daring Romantic speculations about the uncertain future of humanity in the context of emerging new understandings of the changeability of landscapes and species.

Chapter 2 casts a wider net, reflecting the growing mass print culture of the Victorian period in which popular distillations of scientific knowledge and practical how-to guides for amateur naturalists blurred with aesthetic criticism, religious sermons and personal memoirs. We argue that, in parallel to Emerson's and Thoreau's spiritualisations of natural environments, authors in Victorian Britain infused the living world with metaphysical value and meaning that could only be unlocked by combining empirical attentiveness with intuitive feelings. We designate this attitude *conservative idealism*, indicating less a consistent politics than a general rejection of mechanistic materialism and utilitarian rationalism (a sentiment that was of a piece with the evangelicalism that dominated Victorian middle-class culture).[73] This perspective is traced through the art critic John Ruskin's sprawling meditations on the beauty of natural objects. We contend that, by stressing nature's strangeness and unknowability, Ruskin located knowledge of the natural world in mysterious, epiphanic revelations as well as methodical induction. While for authors like Hugh Miller and Charles Kingsley, this revelatory approach to nature reinforced Christian orthodoxy, Richard Jefferies deployed it to construct an idiosyncratic pagan mysticism. It is argued that the revelatory mode cohered with a tendency to deprecate literature's efforts to capture nature's essences. Ruskin, Jefferies, Margaret Gatty, Philip Henry Gosse and William Henry Hudson are all shown to associate authentic experiences of nature with indescribability. Their scepticism of ecomimesis was symbiotic with the rambling narrative and essayistic styles they adopted, offering impressions of nature's variety and interconnectedness instead of seeking to represent it systematically.

The holism of conservative idealism also evoked ancient notions of self-culture, framing nature studies as a form of liberal education. Examining writings by Miller, Kingsley and Gosse, we suggest that these authors depicted this education as simultaneously work and leisure. The tension between the georgic and pastoral dovetailed with underlying tensions between aristocratic and bourgeois values. While Miller presented his work as a stonemason as the foundation of his passion for natural history, Hudson and Jefferies contrasted the leisurely nature appreciator with the toiling labourers who supposedly lacked the former's mental distance from the landscape. We further show that conservative idealism sometimes aligned with inchoate ecological consciousness and trans-species sympathy. Ruskin, Jefferies and Hudson defended the preservation of natural environments and wildlife on the basis that they were fundamental to humans' mental-spiritual well-being. Similarly, William Warde Fowler and Eliza Brighten associated their nature studies with the cultivation of benevolence. Peacefully observing wild animals, they suggested, was to extend one's circles of fellow feeling. We conclude that these ecological and animal-liberationist tendencies were constrained, however, by the inevitable anthropocentrism of valuing the natural world as a moral-spiritual aid.

The discussion then moves onto writing that emphasised humanity's estrangement and distance from 'wild', nonhuman nature. Miller, Gosse and Jefferies are shown to withdraw from some natural phenomena in

[73] See Boyd Hilton, *The Age of Atonement: The Influence of Evangelicalism on Social and Economic Thought, 1795–1865* (Oxford: Oxford University Press, 1988), p. 8.

horror, stressing that nature does not necessarily offer a straightforward model for human life or treasury of wisdom. We further trace a sense of humans' alienation from the rest of the natural world through Leslie Stephen's essays on mountaineering, which envisage natural environments as constructs of the literary imagination. At the same time, analysis of Jefferies' and Hudson's writings reveals dissatisfaction with simplistic anthropomorphism as both authors stress the probable alterity of nonhuman experience. Finally, the chapter considers how Victorian nature writing linked fauna, flora and landscapes to racial, national and regional identities. We find Gatty's, Kingsley's and Jefferies' descriptions of wildlife policing nativeness and foreignness, seeking a discrete and unitary Englishness or Britishness beneath biogeographical variations and continua. Jefferies particularly envisages the national heritage of literature and folklore as an organic growth bound up with specific animal and plant life. Conversely, Hudson is shown to problematise the native–foreign binary in terms of both wildlife and the history of human responses to it. The discussion further explores how Victorian nature writing sometimes undermined the rural–urban binary. Brightwen's, Gosse's and Jefferies' circuitous texts are found to pursue their wild subjects into human homes, gardens and city streets, while Jefferies imagines agricultural machinery blending with the organic processes of the landscape.

Chapter 3 addresses the Modern period. As in other chapters, this period is not seen in strict historical terms and the authors involved in it, whose work is itself generically eclectic, range from Edward Thomas, who was killed by a shell blast towards the end of the First World War (1917), to Richard Mabey and Jane Goodall, who are very much alive and well today (2021). As good a way as any of designating the period is via Eric Hobsbawm's moniker the Short Twentieth Century (1914–91), which he sees as an age of extremes marked, especially in its first half, by a sequence of potentially world-ending catastrophes bookended by the two world wars. Much nature writing of the time evidences this apocalyptic trend, though it also takes heart in the regenerative properties of nature, which is viewed, via the lenses of such fast-developing disciplines as ecology and ethology, with an increasingly scientific eye. This 'observational ethic' – to borrow Caroline Hovanec's resonant term – is first explored in the early twentieth-century fieldwork-based accounts of the prominent ornithologist E. M. Nicholson and the pioneering ecologist Charles Elton, then applied to the post-Darwinian animal writings of D. H. Lawrence and Virginia Woolf and, fast-forwarding several decades, to the popular conservationist vehicles of Gerald Durrell and Jane Goodall, which consider the possibilities of human–animal kinship and the protectionist measures needed to combat species extinction in an ecologically threatened world.

These ecological threats are then brought to bear on early- to mid-twentieth-century rural writing, which is often all too hastily viewed as reacting against the modern forces of industrialisation and urbanisation, but which is better seen as belonging to a complex machinery of rural representation adapted to the cultural needs of post-war England as well as the new technological demands of a rapidly modernising world. Such writing, although it draws on some of the transhistorical devices and tropes commonly associated with nature writing (such as those attached to the conventional modes of the pastoral and the georgic), is much more adept than it is often given credit for in registering the social and environmental changes of its times. These changes can be indexed to increasing anxieties in the national consciousness, while rural writers of the time, far from being monopolised by the English, exhibited a variety of regional/national sensibilities more attuned to the archipelagic context of the British Isles. We look, accordingly, at the work of Edward Thomas (who was of Welsh ancestry) and Margiad Evans (who focused on the Welsh borders), at the work of Irish-born John Stewart Collis, and at the work of Nan Shepherd, whose classic nature study *The Living Mountain*, one of the few texts of its kind to be explicitly modernist in its style as well as modern in its consciousness, has done more than any other to bring to life the rugged mountain landscapes of Scotland's Cairngorms.

What counts as rural writing is moot, as for that matter is what counts as nature writing, and while several writers of the period looked to identify themselves with either the British or the American traditions of nature-themed writing, there was little consensus on what nature writing's main objectives were or, indeed, on what the term 'nature writing' – whose provenance and purpose continue to be disputed – actually meant. At the same time, and counter-intuitively perhaps, the Modern period witnessed the emergence of the genre most of us would immediately recognise today as nature writing: that self-consciously modern, predominantly secular form of stylised first-person narrative that weaves between autobiographical self-awareness, scientific inquiry and environmental concern. The jury remains out, however, as to whether this is the most authentic or effective version of the form, and as this chapter among others makes clear, these debates were by no means restricted to any one historical period, even if the volume surrounding them has recently been turned up. It seems appropriate, then, that the chapter ends with some short reflections on a contemporary nature writer, Richard Mabey, who – rather like William Henry Hudson at the turn of the twentieth century – acts as something of a hinge figure, connecting the concerns of the Modern period to our current times.

Chapter 4 then turns its attention more fully to contemporary nature writing, initially focusing on the so-called new nature writing of the past few decades, but also gesturing back to earlier authors such as Henry Williamson, J. A. Baker and Ronald Blythe. These overlaps, also apparent in Chapter 3, are deliberate insofar as some (if not necessarily all) of these earlier writers share the same concerns about modernity and the violence done in its name as their more recent counterparts; they also suggest that the new nature writing may not be as new, nor indeed as environmentally effective as some of its advocates and practitioners like to claim. On the contrary, as we argue in this chapter, the new nature writing, along with much of the writing nominally associated with the contemporary period, is belated, haunted by a feeling of inadequacy in relation to its predecessors and characterised by the frustrations (lack of reconciliation, failure of transcendence and so on) that Edward Said identifies as features of 'late style'.[74] Many British writers of the period, we contend further, are best seen as 'late Moderns', expressing deep-seated anxieties about themselves, their writing and their position in a rapidly diminishing natural world.

This thesis is examined in relation to three writers – Robert Macfarlane, Jay Griffiths and George Monbiot – whose work simultaneously attempts to recall the wild and reflects on the impossibility of that exercise. Other writers (Cowen, Farley, Roberts) are then brought in to examine those contemporary post-industrial landscapes that might create the conditions of possibility for a 'new wild'. The second half of the chapter pursues this line of argument, but in relation to another popular subgenre, animal writing, which is seen as containing regenerative potential but also as communicating unsettling insights into the always unstable relationship between animal others and human selves. Two groups of species are picked out for closer scrutiny: hawks, whose laser-like vision is a source of both wonder and terror (Macdonald, Baker, Jamie), and otters, who are both objects of violence and violent subjects in themselves (Williamson, Maxwell, Darlington, Foster).

A different type of violence is explored in the chapter's concluding section: the violence of the elements, which – in today's era of accelerated climate change – is both significantly influenced by human beings and

[74] Edward W. Said, *On Late Style* (London: Bloomsbury, 2006).

beyond the bounds of human control. Taking in works by Melissa Harrison and Richard Mabey, we gauge that most reassuring of British pursuits, conversation about the weather, against 'weather-worlds' (Ingold) in which the elements are constantly in flux, with consequences that are anything but reassuring.[75] As Harrison, Mabey and, indeed, most of the writers featured in this chapter show, contemporary considerations of weather and climate must contend with the fact that we live in a 'post-natural' world in which previously taken-for-granted understandings of 'nature' and 'the natural' are scarcely viable. However, this is by no means a new phenomenon, as is made apparent in this book's earlier chapters. Rather, it offers a reminder that the modern world to which nature writing speaks is inherently conflicted, and that nature writing – from the Romantic period right through to the present – is not the calming refuge it is still sometimes mistaken for, but a source of multiplying unease: a multifaceted registration of continual threat.

[75] Timothy Ingold, *Being Alive: Essays on Movement, Knowledge and Description* (London: Routledge, 2011), pp. 96, 131–2.

CHAPTER I

Romantic

Introduction

The story of Romantic nature writing has been told many times. Perhaps the longest-standing narrative is that Romantic poets, reacting to the Enlightenment's rationalistic and mechanistic approach to the more-than-human world, found in nature a quasi-religious sense of communion. Despite the huge body of scholarship revealing the complexity and heterogeneity of both the Enlightenment and Romanticism, this narrative continues to resonate. It has many variants. Critics have celebrated and lamented how canonical figures like William Wordsworth and Samuel Taylor Coleridge transformed their early hopes for political revolution into visions of individual apotheosis aided by the natural sublime.[1] Others have suggested that when the Romantics claim to be writing about Nature, they are really writing about History.[2] More recently, scholars have shown how emerging fields of knowledge such as geology influenced Romantic literature and vice versa.[3] Important work has been done on the complex relationship between culture, natural science, exploration, race and empire.[4] Romantic writers have been celebrated for articulating a new ecological consciousness resisting modernising forces such as industrialisation

[1] M. H. Abrams, *Natural Supernaturalism* (New York: Norton, 1973); Jerome McGann, *The Romantic Ideology* (Chicago, IL: University of Chicago Press, 1983).

[2] Alan Liu, *Wordsworth: The Sense of History* (Stanford, CA: Stanford University Press, 1989).

[3] Noah Heringman, *Romantic Rocks, Aesthetic Geology* (Ithaca, NY: Cornell University Press, 2004); Ralph O'Connor, *The Earth on Show* (Chicago, IL: University of Chicago Press, 2007).

[4] See, for example, Alan Bewell, *Natures in Translation* (Baltimore, MD: Johns Hopkins University Press, 2017); Tim Fulford, Debbie Lee and Peter Kitson, *Literature, Science and Exploration in the Romantic Period* (Cambridge: Cambridge University Press, 2004); Kevin Hutchings, *Romantic Ecologies and Colonial Cultures in the British-Atlantic World, 1770–1850* (Montreal: McGill-Queen's University Press, 2009).

and enclosure.[5] They have also been criticised for valorising 'the egotistical sublime', with its legacy in contemporary nature writing of the figure of the 'lone enraptured male' colonising wild space.[6]

All these stories are useful and not without their truths, but this chapter takes a different approach. What might the story of Romantic nature writing look like if its focus was on prose rather than poetry? How did prose writers mediate between specific natural-historical knowledge and broader theological-philosophical ideas about nature? How did they balance self-expression with external observation? What can analysis of the formal properties of Romantic nature writing tell us about the subsequent history of the genre? To what extent does it even make sense to discuss 'nature writing' as a discrete form in the Romantic period? The story we uncover has its limitations, like all the others, but it at least gives the lie to any lingering notions of Romantic nature writing as naïve or idealised. What emerges instead is its complexity, its rhetorical self-consciousness and even its modernity. The most interesting nature writing tends to be self-reflexive: 'as much an interrogation as a performance of mimesis', as we put it in the Introduction.[7] To suggest that Romantic nature writing is modern is not, however, to efface its historicity. Rather, it is to understand it genealogically and to attend to its urgent engagement with the various crises of modernity: environmental, political, ontological, epistemological and representational. It is also to draw attention to how Romantic nature writing self-consciously engages with the future as well as the past.

Bruno Latour's work offers a helpful way of thinking about this issue, for Romantic nature writers were often reacting to what he identifies as 'the modern constitution' – that is, a mode of thinking that separates nature and culture into separate ontological realms and, by doing so, allows for a proliferation of hybrids that is simultaneously denied.[8] Romantic nature writing often pushes against seeing something called 'nature' as a mere object of study and human mastery, instead representing an emerging and evolving set of relations that includes humans and nonhumans: an

5 For an excellent overview of the field, see Jeremy Davies, 'Romantic Ecocriticism: History and Prospects', *Literature Compass*, 15 (2018).

6 See Kathleen Jamie 's review of Robert Macfarlane's *The Wild Places* (2007), *London Review of Books*, 30 (6 March 2008), and also Jason Cowley's contrast between old nature writing – 'the lyrical pastoral tradition of the romantic wanderer' – and a supposedly more 'heterodox and experimental' new form: 'The New Nature Writing', *Granta*, 102 (Summer 2008), p. 10.

7 For the limitations of ecomimesis, see Timothy Morton, *Ecology without Nature* (Cambridge, MA: Harvard University Press, 2009).

8 Bruno Latour, *We Have Never Been Modern*, trans. Catherine Porter (Cambridge, MA: Harvard University Press, 1993).

ontology based on becoming rather than being. Its modernity, therefore, might be seen in its resistance to those other elements of modernity that seek separation rather than entanglement: a resistance that looks both forwards and backwards. Indeed, Romantic writers often seem to have something in common with recent new materialist thinkers such as Karen Barad and Jane Bennett.[9] For example, Barad suggests that 'the primary ontological unit is not independent objects with inherent boundaries and properties but rather *phenomena*'.[10] Objects and agents emerge through mutual relations (hence her use of the term 'intra-action') rather than acting as pre-existing entities. Similarly, the writers discussed in this book tend to be concerned not with the natural world as a mere backdrop to human activity, or as a set of reified taxonomic categories, but with nature in motion: a rich assemblage of organisms and phenomena.

Historical Contexts

The historical contexts for Romantic nature writing are so complex that one could easily get lost in them. What follows can be no more than a simplified schema. For heuristic purposes, the list presented here splits those contexts into epistemological, sociopolitical and literary: elements which were in practice closely related. Lacking the space to provide detailed discussion, we have instead provided references to key scholarship.

Epistemological[11]

1. The new idea of 'deep time' and (concomitantly) the contingency of humanity
2. The development of universal taxonomy (Linnaeus)

9 See Evan Gottlieb, *Romantic Realities: Speculative Realism and British Romanticism* (Edinburgh: Edinburgh University Press, 2016). One key reason for this commonality is that British Romantic writers are generally working within a pre-Kantian philosophical paradigm, and new materialism reacts to post-Kantian anthropocentrism. Some of the more philosophically minded Romantics (e.g. Coleridge) were highly influenced by Spinoza's ideas, which can seem close to key aspects of the new materialism.

10 Karen Barad, *Meeting the Universe Halfway: Quantum Physics and the Entanglement of Matter and Meaning* (Durham, NC: Duke University Press, 2008), p. 139.

11 Melissa Bailes, *Questioning Nature: British Women's Scientific Writing and Literary Originality, 1750–1830* (Charlottesville: University of Virginia Press, 2017); Mary Ellen Bellanca, *Daybooks of Discovery: Nature Diaries in Britain, 1770–1870* (Charlottesville: University of Virginia Press, 2007); Hannah Hodacs, Kenneth Nyberg and Stéphane van Damme (eds.), *Linnaeus, Natural History and the Circulation of Knowledge* (Oxford: Voltaire Foundation, 2018); Peter H. Reill, *Vitalizing Nature in the Enlightenment* (Berkeley: University of California Press, 2005); Martin Rudwick, *Bursting the Limits of Time: The Reconstruction of Geohistory in the Age of Revolution* (Chicago, IL: University of

3. Broad trends in natural philosophy and natural history, including the early stirrings of disciplinary formations such as phenology, ethology and ecology
4. Philosophical accounts of 'Nature' not as brute matter but as a dynamic set of processes imbued with spirit (e.g. in German *Naturphilosophie* and, going back further, Spinoza)

Sociopolitical[12]

1. Empire, colonialism and the slave trade
2. Enclosure and changing patterns of agricultural land use
3. The emergence of industrial capitalism
4. Revolutionary politics (often represented by figures of environmental catastrophe)
5. Prevalent cultural attitudes to nature (e.g. blood sports, egg collecting) and resistance to them

Literary[13]

1. Colonial travel narratives
2. Antiquarian texts (local histories, ballad collections)
3. Reworking of classical poetic genres (especially pastoral and georgic)
4. Writings on agricultural 'improvement'
5. The growth of autobiography as a secular genre

These contexts are all important, but imperialism and colonialism are crucial to an understanding of Romantic nature writing. They drove the development of Western natural history as a science and the generation of

Chicago Press, 2005); Paul Warde, *The Invention of Sustainability* (Cambridge: Cambridge University Press, 2018).

[12] Peter Heymans, *Animality in British Romanticism: The Aesthetics of Species* (New York: Routledge, 2012); Donna Landry, *The Invention of the Countryside* (Basingstoke: Palgrave, 2001); Andreas Malm, *Fossil Capital* (New York: Verso, 2016); Mary Ashburn Miller, *A Natural History of Revolution* (Ithaca, NY: Cornell University Press, 2011).

[13] Kevis Goodman, *Georgic Modernity and British Romanticism* (Cambridge: Cambridge University Press, 2004); Noah Heringman, *Sciences of Antiquity: Romantic Antiquarianism, Natural History, and Knowledge Work* (Oxford: Oxford University Press, 2013); Nigel Leask, *Curiosity and the Aesthetics of Travel Writing, 1770–1840* (Oxford: Oxford University Press, 2002) and *Robert Burns and Pastoral: Poetry and Improvement in Late Eighteenth-Century Scotland* (Oxford: Oxford University Press, 2010).

universal taxonomy.[14] Even the most locally focused British Romantic writers wrote self-consciously within a transnational context shaped by significant global circulations of plants, animals, people, ideas and texts.[15] The most important and appalling of these circulations was the transatlantic slave trade. Romantic natural history was inevitably caught up with racialised thinking and, at times, unambiguous articulations of white supremacy, most obviously through the assertion of hierarchies of race.[16] These articulations are rarely found in the kind of personal nature writing we discuss in this chapter, with the exception of William Cobbett's works, but that absence does not mean our texts are racially egalitarian. The supremacy of European culture and whiteness itself was so ingrained in Romantic-period discourse that it often went without saying.[17] Despite the revolutionary politics of the period, Romantic nature writing was also inflected by hierarchies of class and gender from which even the most liberal writers could not fully break free. And these hierarchies were themselves inflected by global phenomena. For example, empire significantly changed land usage and access within the British Isles, through the capital accumulation it enabled among the fortunate few and the stripping away of common rights from the unfortunate many, including John Clare (one of our key authors), who were badly affected by enclosure and engrossing (the merging of multiple small farms). We do not claim that colonial land theft and enclosure within Britain were commensurate, but certainly they were inextricably linked, in part because they often involved the privatisation of commonly held land for the explicit purpose of 'improvement', and generally for the benefit of an elite. The displacement of rural populations within Britain also gave industrialists access to the large urban workforces that would be crucial to Britain's emergence as a world power.[18]

[14] Subhadra Das and Miranda Lowe, 'Nature Read in Black and White: Decolonial Approaches to Interpreting Natural History Collections', *Journal of Natural Science Collections*, 6 (2018), 4–14; Gyan Prakash, 'Science "Gone Native" in Colonial India', *Representations*, 40 (1992), 153–78.

[15] See Bewell, *Natures in Translation*, and David Higgins, *Romantic Englishness: Local, National, and Global Selves, 1780–1850* (London: Palgrave, 2014).

[16] B. Ricardo Brown, *Until Darwin: Science, Human Variety, and the Origins of Race* (London: Pickering and Chatto, 2010).

[17] This structural racism has often been replicated in the subsequent academic study of Romanticism; for a powerful challenge, see the Bigger 6 Romanticism project: bigger6romantix.squarespace.com/ (accessed 1 September 2020). Recent studies challenging Eurocentric constructions of Romanticism include Manu Chander, *Brown Romantics: Poetry and Nationalism in the Global Nineteenth Century* (Lewisburg, PA: Bucknell University Press, 2017), and Nikki Hessell, *Romantic Literature and the Colonised World* (London: Palgrave, 2018).

[18] For a more detailed account of the relationship between enclosure, colonialism, capitalism and inequality, see Jason Hickel, *The Divide* (London: Penguin, 2017), pp. 76–89.

Selection of Texts

In what follows, we have taken a selective approach, as we also do in other chapters, and we have emphasised depth rather than range. Much canonical Romantic 'nature poetry' does not feature at all, nor does some popular natural history of the period. We have focused instead on those texts that reveal something important about the development of nature writing as a prose form. Broadly speaking, four key things connect them: (1) a concern with nature in motion as opposed to static taxonomy, (2) formal complexity, (3) accounts based on localised personal experience and (4) an interest in the troubled entanglements of the human and nonhuman worlds. We begin with a discussion of Gilbert White, whose *The Natural History and Antiquities of Selborne* (1789) is often cited as a foundational text in British nature writing. However, we complicate this standard origin story by looking at White's work alongside the work of Jean-Jacques Rousseau and showing how the latter's confessional mode has also played a vital role in the development of the form. We then find in Dorothy Wordsworth's journals a kind of middle way between Rousseau's self-exploration and White's self-effacement, although we also discuss significant similarities in how all three writers emphasise the dynamism of nonhuman nature. The following section addresses two figures whose contribution to British nature writing has yet fully to be acknowledged: the protean author Charlotte Smith and the natural historian and wood engraver Thomas Bewick. Both used experimental forms – for Smith the interplay of prose and poetry, and for Bewick that of text and image – in order to explore not only natural history, but also the metonymic connections between nature and culture. Our final section concerns two labouring-class writers: James Hogg and John Clare. Both resisted and lamented modernity, but did so through developing new modes of writing that appealed to self-consciously modern and urban readership. Like all the best nature writing, their texts combine grounded observation of nonhuman nature with a formal sophistication, even a playfulness, that addresses the fraught relationship between text and world.

Before we begin, it will be helpful to explain our omissions. For reasons of feasibility rather than parochialism, this chapter does not address colonial travel writing. The writers featured are also all white. There is a small but important body of black British writing in the Romantic period, most notably in the form of the 'slave narratives' of authors such as Olaudah Equiano, Mary Prince and Robert Wedderburn, but we have not found any that would plausibly qualify as nature writing.[19] The role of British nature writing in constructions of race and nation is vital, however, and something we return to throughout this book. There is also the issue of Britishness. For a long time, critics tended to conflate British and English Romantic literature. In recent years, a much richer understanding of 'four nations' Romanticism has emerged. However, while writers from Wales, Scotland and Ireland had an enormous impact on most genres, their impact on the kind of personal nature writing examined in this chapter is less palpable. With the obvious exceptions of Rousseau and Hogg, our main authors are English. An extended survey of the varieties of British nature writing in the period would be a valuable if challenging project and would offer a more capacious 'four nations' approach. It could take in the earlier natural history of John Ray (English) and William Withering (English); the creative antiquarianism of James Macpherson (Scottish), Iolo Morganwg (Welsh) and Thomas Percy (English); the innovative geological writings of William Buckland (English), James Hutton (Scottish) and William Smith (English); the agricultural tracts of William Cobbett (English) and Arthur Young (English); the poetic natural philosophy of Erasmus Darwin (English); the scientific writings of Humphry Davy (English) and Jane Marcet (English/Swiss); the natural histories of James Dickson (Scottish), Elizabeth Kent (English), John Richardson (Scottish) and Priscilla Wakefield (English); and the poetry of Anna Laetitia Barbauld (English), Robert Bloomfield (English), Lord Byron (Anglo-Scottish), Robert Burns (Scottish), Samuel Taylor Coleridge (English), Allan Cunningham (Scottish), Thomas Moore (Irish), Allan Ramsay (Scottish), Mary Robinson (English), Anna Seward (English), Percy Bysshe Shelley (English) and William Wordsworth (English). But, like the rest of this book, this chapter does not offer a survey. Rather, it seeks to bring out the most important Romantic influences in the genealogy of modern British nature writing in order to understand what is distinctive and important about the form.

[19] See Ryan Hanley, *Beyond Slavery and Abolition: Black British Writing, c. 1770–1830* (Cambridge: Cambridge University Press, 2019); Helen Thomas, *Romanticism and Slave Narratives* (Cambridge: Cambridge University Press, 2000). There is no doubt either that indigenous people made significant unacknowledged contributions to European colonial natural history; however, unpicking those contributions, were it possible, is beyond the scope of this chapter.

Ecomimesis and the Self: Jean-Jacques Rousseau, Gilbert White and Dorothy Wordsworth

Modern British nature writing owes a great deal to Gilbert White. *The Natural History of Selborne* (1789) presented nonhuman creatures as forming a dynamic ecological community. Its sympathetic and precise attention to animal behaviour, as well as its close focus on a particular locale, have had a powerful influence on nature writers over the past two centuries. However, *Selborne* differs in one very important respect from many of the most popular recent nature books, such as Richard Mabey's *Nature Cure* (2005), Helen Macdonald's *H Is for Hawk* (2014) or Amy Liptrot's *The Outrun* (2016). It may be personal, but it is rarely intimate. White actually shows relatively little interest in what encountering nonhuman nature *does* to the self. Contemporary nature writing is much more concerned with this topic and often contains a confessional element. This is true even of works that do not offer themselves as recovery narratives per se. Immersion in nature (however conceived) is usually presented, if not as a straightforward cure for psychological ailments, then at least as potentially therapeutic, and it is common for nature writers to include autobiographical detail, such as family relationships, personal problems and traumas, and childhood experiences. The causes of this trend are no doubt overdetermined; among other things, it reflects a separation that had not yet fully formed in the Romantic period between nature writing and natural history as science (see the Introduction and Chapter 2 for further discussion).

The Romantic paradigm for autobiographical nature writing was offered not by Gilbert White, but by Jean-Jacques Rousseau, whose confessional prose revolutionised the potential for self-expression in European literature. Rousseau was 'willing to lay bare his psychological fragility and human vulnerability' to an extent that was both shocking and exhilarating for eighteenth-century readers.[20] His autobiographical texts, particularly *Confessions* (1782–9) and *Les Rêveries du promeneur solitaire* (1782), were central to his broader philosophical and political project in that they sought to harness the republican potential of self-revelation and to present alienation from nature as key to the ills of modern life. Rousseau's turn to confession later in life coincided with his development of an obsession with botany, which he was to explore in depth in 'Seventh Walk' from *Rêveries*. Like *Selborne*, Rousseau's essay emphasises

[20] Jean-Jacques Rousseau, *Reveries of the Solitary Walker*, ed. and trans.' Russell Goulbourne (Oxford: Oxford University Press, 2011), p. xxviii. Further references are to this edition and are included in the text.

first-person experience and fieldwork rather than abstract taxonomy. But Rousseau's essay also has three connected elements that one might expect to find in the work of present-day nature writers. The first of these elements is a detailed account of the writer's self. As Russell Goulbourne points out, *Rêveries* does not offer the 'moral vivisection' of the *Confessions* (p. xiii); however, it nonetheless promises the reader access to the author's innermost thoughts and feelings and bears the scars of self-doubt and paranoia. The second element is an emphasis on the therapeutic potential of encounters with nonhuman nature. Botanising walks, for Rousseau, offer 'recreation for the eyes, which in misfortune relaxes, amuses, and distracts the mind and lifts the feelings of pain' (p. 72). He understands botany as a form of *askesis*: a treatment for the soul's maladies.[21] The final element is formal self-consciousness: an awareness of the contradictions of writing nature, the irony of the writer's own subject position, and the limitations of imagining nature as separate from culture.

All three elements are encapsulated by a passage from 'Seventh Walk', in which Rousseau describes a solitary botany expedition he undertook in the canton of Jura in Switzerland:

> I went deep along the winding paths up the mountain, and ... I finally reached a refuge that was so hidden that it was wilder than anything I have ever seen in my life. Black fir trees ... formed an impenetrable barrier around this refuge; all that could be seen through the few gaps in this dark wall was sheer rock faces and terrifying chasms which I only dared look into while lying flat on my stomach. (p. 79)

This passage reflects what was in the late eighteenth century a relatively new celebration of mountain landscapes. Rousseau's description seems exemplary of the sublime mode, presenting the solitary individual escaping from social life into superlatively 'wild' space: an experience that is both frightening and enthralling. For contemporary nature writers and ecocritics, such invocations of the sublime are to be treated with scepticism, in part because of the inequalities of class, race and gender they occlude (see Chapter 4 and Afterword). The reflections prompted by Rousseau's walk seem almost designed to trigger ideological alarm bells. 'I compared myself to those great travellers who discover a desert island, and I said complacently to myself: "I am undoubtedly the first mortal ever to have reached this place"; I considered myself a second Columbus' (p. 80). At first glance, the passage represents the *ne plus ultra* of the imperialising sublime male for

[21] Alexandra Cook, *Jean-Jacques Rousseau and Botany* (Oxford: Voltaire Foundation, 2012), p. 2.

which Romanticism is so often criticised. After all, Rousseau here imagines himself as an explorer and a conqueror of apparently untouched space, and compares himself to a figure now associated with colonial genocide.

However, the passage quickly becomes more complicated. First, there is a strong sense of connection to the flora and fauna around the autobiographical subject that is not always present in later Rousseauvian evocations of the mountain sublime (such as Canto III of Byron's *Childe Harold's Pilgrimage* (1816)). 'From mountain crevices', Rousseau hears the cries of 'the horned owl, the little owl, and barn owl', and finds 'the horror' of solitude lightened by 'a few rare but familiar little birds' (p. 79). He also records with botanical exactitude that he encountered 'seven-leaved toothwort, cyclamen, *nidus avis*, the large *laserpitum*' (p. 80). Even as he falls into a 'reverie ... imagining that I was in a refuge unknown to the whole universe' and forgets about 'botany and plants', he notes that he was 'sitting on cushions of *lycopodium* and mosses' (p. 80). The hint here of a split between the self who experiences sublime reverie and the more observant narrator who views the scene in retrospect is crucial to understanding *Rêveries*, as it is to other Rousseauvian autobiographies such as William Wordsworth's *The Prelude* (1850). The split also appears earlier in the passage, when Rousseau states that he 'complacently' imagined himself as a 'second Columbus' (p. 80). The rapturous self who goes on the walk is ironised, and the gap opens up further as the passage continues:

> While I was strutting about ... I heard not far off a certain clicking noise which I thought I recognized ... Surprised and intrigued, I got up, pushed my way through a thicket of undergrowth in the direction of the noise, and, set in a little valley twenty yards from the place that I thought I was the first person to have reached, I saw a stocking factory. (p. 80)

Like 'complacently', 'strutting' suggests self-mockery. As Rousseau describes it, his discovery of the penetration of human industry into the supposedly wildest of spaces elicits comically contradictory emotions. Initial 'joy' at the prospect of company is followed by 'a more lasting feeling of distress at not being able, even in the caves of the Alps, to escape the cruel clutches of those men bent on tormenting me' (p. 80). There is no doubt that Rousseau experienced genuine persecution in the latter stages of his life, but he also suffered from the severe paranoia that is evident here. However, if his apparent escape from society into the wilderness has proved to be a mirage, it has allowed him to shift into a healthier perspective. At first, he is 'absolutely sure' that the workers are co-conspirators in a plot orchestrated by an enemy. But he quickly moves into a different view: 'I ended up laughing to myself both because of my childish vanity and because of the comic way in which I had been punished for it' (p. 80).

Rousseau begins by emphasising the sublimity of wilderness experience in a way that now seems archetypically 'Romantic' but in the late eighteenth century could still seem innovative. The passage develops into a fluid and self-aware reflection of the limitations of its own investment in that sublimity. It ends with the kind of ironic juxtaposition of human activity and 'wild' space that is characteristic of nature writing in the Anthropocene (see Chapter 4). The rest of 'Seventh Walk' is also rhetorically complex. Rousseau presents his senescent obsession with botany self-critically, as an 'enthusiasm that smacks of extravagance' and a 'madness' (p. 69). He pronounces polemically on the deadening instrumentality of seeing plants only as pharmacological resources or as a means to becoming 'an author or a professor' (p. 78). And he musters semi-comic hyperbole in describing the horrors of investigating the 'mineral kingdom' (p. 75) – evoking 'the haggard faces of the wretches languishing in the foul vapours of the mine' (p. 76) – or of studying animals in an 'anatomy theatre': 'stinking corpses, oozing and livid flesh, blood, disgusting intestines, awful skeletons, and pestilential fumes! Believe me, that is not the place where J.-J. will go looking for amusement' (p. 77). His distaste for the gruesome theatricality of dissection – typical of his general suspicion of the theatre – is complicated by his own rather stagey imagining of the scene, including the appearance of a version of himself performing that distaste.

Rousseau presents botany as superior because it encourages fieldwork in the outdoors, does not require the dismemberment of specimens, and does not necessitate social performance. Its genre is not the theatre, but a particular kind of book: the herbarium. As Alexandra Cook has shown, herbaria – collections of preserved plants mounted in books – were a 'well-established part of botany when Rousseau took up plant studies'.[22] Their preparation required considerable skill: plants had to be identified, gathered, dried, preserved and mounted. Rousseau followed Linnaeus in presenting 'his specimens one to a page, following known principles of composition'.[23] However, he also innovated, particularly in creating smaller *herbiers portatifs* that were more suitable for popularising botany than the standard folio size.[24] On the final page of 'Seventh Walk',

[22] Cook, *Jean-Jacques Rousseau and Botany*, p. 255. [23] *Ibid.*, p. 266. [24] *Ibid.*, pp. 290–2.

Rousseau evokes the capacity of the herbarium to function as a *lieu de mémoire*:

> Now that I can no longer roam about those glorious places, all I have to do is open my herbarium and it quickly transports me there. The pieces of plants that I gathered there are enough to remind me of the whole magnificent spectacle. This herbarium is for me a diary of my botanical expeditions which makes me set off on them again with renewed delight and which produces the effect of an optical chamber, showing them again before my very eyes. (p. 82)

If modern nature writing is frequently troubled by its own textuality – the problem being that words may alienate us from nonhuman nature even as they connect us – then the herbarium offers a potential solution. Unlike a book of nature writing, the herbarium is a hybrid form, combining as it does human technology (the book itself; chemicals used for preserving and mounting), forms of mediation (language used to classify specimens; principles of visual composition) and the plants themselves. More so than other forms of nature writing, it might therefore exemplify Gilles Deleuze and Félix Guattari's claim that 'the book is not an image of the world. It forms a rhizome with the world.'[25] That is, the herbarium functions as an assemblage rather than as a representation. If, as Jacques Derrida has famously argued, Rousseau had a deep distrust of writing (predicated on absence) in contrast to the supposedly more natural intimacy of speech (predicated on presence), then the herbarium may have seemed a more effective form of ecomimesis.[26] By bringing the reader into direct contact with 'pieces of plants', it enables him or her to be transported through memory into a living ecosystem.

Rousseau was well known and much read in Britain even before he became notorious in the 1790s as one of the principal architects of the French Revolution. The French edition of *Rêveries* was reviewed by British journals when it was published – reading French was a common accomplishment among the British elite – and English translations followed in 1783 and 1790. As many critics have shown, Rousseau profoundly influenced British Romantic literature.[27] He was also commonly criticised, particularly by counter-revolutionary thinkers like Edmund Burke. That

[25] Gilles Deleuze and Félix Guattari, *A Thousand Plateaus: Capitalism and Schizophrenia*, trans. Brian Massumi (London: Bloomsbury, 2016), p. 10.

[26] Jacques Derrida, *Of Grammatology*, trans. Gayatri Chakravorty Spivak (Baltimore, MD: Johns Hopkins University Press, 1998).

[27] Russell Goulbourne and David Higgins (eds.), *Jean-Jacques Rousseau and British Romanticism* (London: Bloomsbury, 2017).

modern British nature writing would not acknowledge his influence is hardly surprising, given his association with Romanticism's worst excesses. Among other things, Burke's effective hatchet job presented him as exemplary of a dangerous Continental tendency to abstract thought in contrast to a healthy English empiricism. But Rousseau was in fact a knowledgeable naturalist, and self-reflexive on nature writing as a literary form. If Romanticism has often seemed like a problem to modern nature writers, this is perhaps because its combination of groundedness and rhetorical complexity has too often been forgotten or strategically repressed.

Unlike Rousseau, Gilbert White is often celebrated for his innovative depiction of a 'living nature', as opposed to 'reproducing the textual equivalent of a library or museum'.[28] He takes full advantage of the opportunities provided by *Selborne*'s epistolary form, offering lively vignettes and personal reflections. Despite its narrow geographical focus, it provides a great variety of observations and narratives, a variety enabled by the familiarity and occasionality of its letters. Although he was well versed in Linnaean nomenclature and the works of taxonomic natural historians, White self-consciously distanced himself from that approach:

> Faunists, as you observe, are too apt to acquiesce in bare descriptions, and a few synonyms: the reason is plain; because all that may be done at home in a man's study, but the investigation of the life and *conversation* of animals, is a concern of much more trouble and difficulty, and is not to be attained but by the active and inquisitive, and by those that reside much in the country.[29] (emphasis mine; see also Chapter 3)

White is using the term 'conversation' in a now archaic sense to mean all kinds of interactions.[30] His emphasis on understanding the dynamic processes of nonhuman nature, based on empirical experience, has offered a powerful legacy to modern nature writers. Also influential has been his related emphasis on the value of '*parochial history*', not only of human-made 'antiquities', but also of 'natural productions and occurrences' (p. 3). On the face of it, White's localism might seem in opposition to the

[28] Bewell, *Natures in Translation*, p. 155. White's originality can be overstated. For useful contextualisation, see Brycchan Carey, Sayre Greenfield and Anne Milne (eds.), *Birds in Eighteenth-Century Literature* (London: Palgrave, 2020), especially chapter 9; Noah Heringman, 'Natural History in the Romantic Period', in Jon Klancher (ed.), *A Concise Companion to the Romantic Age* (Oxford: Wiley-Blackwell, 2009), pp. 141–67.

[29] Gilbert White, *The Natural History of Selborne*, ed. Anne Secord (Oxford: Oxford University Press, 2013), p. 120. Further references to *Selborne* are to this edition and are included in the text.

[30] Ted Dadswell, *The Selborne Pioneer: Gilbert White As Naturalist and Scientist* (Farnham: Ashgate, 2003), p. 48.

eighteenth century's grand schemes of imperialistic natural history, as exemplified by the career of Sir Joseph Banks:

> Alongside the emergence of a natural history aimed at collecting, mobilizing, and changing nature on a worldwide scale ... there emerged an alternate, nativist, localist, domestic natural history that valued the familiar and the everyday inhabitants of the English countryside over the exotic and the foreign.[31]

As Alan Bewell and Tobias Menely have shown, these apparently distinct forms of nature writing were really two sides of the same coin. White used similar fieldwork techniques and writing conventions to those of colonial travellers. He also related the natural history of Selborne to global natural history, 'which he came to understand by reading travelogues, examining foreign specimens, comparing notes with other naturalists, and practicing Linnaean taxonomy'.[32] Indeed, White suggests that the work of 'partial writers' is more likely to be accurate than that of generalists, and will thereby 'pave the way to an universal correct natural history' (p. 71).

If the long-standing understanding of White as an exemplary figure of localism needs nuancing, so too does his association with a myth of timeless English landscape. Like most thinkers of his time, he was an adherent of natural theology, which imagined nature as a harmonious and unchanging system reflective of God's benevolence (see Chapter 2). However, this idea was being challenged at the end of the eighteenth century, as it 'did not easily account for changing environments, or for the obvious ecological imbalances and instabilities that were being visibly produced' by globalising forces.[33] White was aware that human activity had caused local extinctions, such as the black grouse and red deer (pp. 16–17), and he also experienced invasive species in Selborne such as the *Blatta orientalis* or Asian cockroach (which actually probably originated in Africa).[34] Stuart Peterfreund identifies a significant tension between White's representation of Selborne as 'the type of an unchanging earthly Paradise' and a sense of 'temporal change' in the apocalyptic language he uses to describe meteorological phenomena.[35] Despite its reputation,

[31] Bewell, *Natures in Translation*, pp. 155–6.
[32] Tobias Menely, 'Traveling in Place: Gilbert White's Cosmopolitan Parochialism', *Eighteenth-Century Life*, 28 (2004), 46–65 (p. 48).
[33] Bewell, *Natures in Translation*, p. 179. [34] *Ibid.*, pp. 189–95.
[35] Stuart Peterfreund, '"Great Frosts and ... Some Very Hot Summers": Strange Weather, the Last Letters, and the Last Days in Gilbert White's *The Natural History of Selborne*', in Noah Heringman (ed.), *Romantic Science: The Literary Forms of Natural History* (Albany: State University of New York Press, 2003), pp. 85–108 (p. 93).

The Natural History of Selborne is not a comfortable book offering a consolatory fantasy of sequestered Englishness, but a complex and in some ways troubled depiction of the dynamism and unpredictability of nonhuman nature, both very much of and equally ahead of its time.

The sheer energy of White's book is encapsulated by a letter of August 1778. 'A good ornithologist', he notes at the start, 'should be able to distinguish between birds by their air as well as by their colours and shape; on the ground as well as on the wing, and in the bush as well as in the hand' (p. 188). White is describing what modern birders call 'jizz': the overall impression ('air') one gets of a bird through its movements and location as well as the more usual taxonomic categories.[36] This process of identification would have been particularly important to naturalists in the late eighteenth century, who did not have access to the powerful binoculars now taken for granted.[37] White suggests that 'Put a bird in motion … *Et vera incessu patuit* … ' (p. 213); the quotation from Virgil's *Aeneid* can be loosely translated as 'and it reveals itself'. The letter itself exhibits a kind of motion, swooping rapidly through kites, buzzards, kestrels, hen harriers, owls, ravens, rooks, crows, jackdaws, woodpeckers, parrots, magpies, jays, herons, pigeons and many other birds. Like later nature writers, White loves to list species and phenomena and generates dynamic prose through parataxis: 'Rooks sometimes dive and tumble in a frolicksome manner; crows and daws swagger in their walk; wood-peckers fly *volatu undoso*, opening and closing their wings at every stroke, and so are always rising and falling in curves' (p. 188).[38] Here the parallel syntax of 'opening and closing' and 'rising and falling' suggests an attempt to capture the woodpecker's movements at the level of style. Different bird species emerge in this letter as having distinctive idiosyncrasies and personalities expressed through flight. Ravens 'spend all their leisure time in striking and cuffing each other on the wing in a kind of playful skirmish'; mistle thrushes 'use a wild and desultory flight'; and the whitethroat 'uses odd jerks and gesticulations over the tops of hedges and bushes' (p. 188–90). As one would expect from a set of notes based on detailed field observations, figurative language is used carefully and sparingly: the greenfinch 'exhibits such languishing and faultering gestures as to appear like a wounded and dying bird'; 'the

36 Brycchan Carey also makes this connection in an illuminating analysis of the letter: see 'The Literary Gilbert White', in Carey et al., *Birds*, pp. 173–92.

37 As Ted Dadswell notes, although he used a telescope to observe a hot-air balloon in October 1784, there is no mention of such a device 'used for bird or animal observation in any of White's extant writings' (*The Selborne Pioneer*, p. 185).

38 *Volatu undosu* means 'in an undulating fashion'.

king-fisher darts along like an arrow'; and 'owls move in a buoyant manner, as if lighter than the air' (pp. 188–90).

The most self-consciously poetic image in the list of 'bird[s] in motion' is of the 'fern-owls or goat-suckers, [which] glance in the dusk over the tops of trees like a meteor' (p. 189). White is referring to the nocturnal species now known as the European nightjar, a subject of popular superstition for supposedly suckling nanny goats and hurting weaning calves, although White dismissed these ideas.[39] The comparison to the 'meteor' (shooting star) is therefore apt, as such phenomena were popularly seen as ominous portents. Perhaps because of the stories attached to it, and its similarity in flight to his beloved hirundines (swifts, swallows and martins), the nightjar was one of White's favourite species: 'a wonderful and curious creature' that he claimed to have studied as carefully as any bird (p. 49). In a 1771 letter, he described its movements in more detail:

> On the twelfth of July I had a fair opportunity of contemplating the motions of the *caprimulgus*, or fern-owl, as it was playing round a large oak that swarmed with *scarabaei solstitiales*, or fern-chafers. The powers of its wing were wonderful, exceeding, if possible, the various evolutions and quick turns of the swallow genus. But the circumstance that pleased me most was, that I saw it distinctly, more than once, put out its short leg while on the wing, and, by a bend of the head, deliver somewhat into its mouth. If it takes any part of its prey with its foot, as I have now the greatest reason to suppose it does these chafers, I no longer wonder at the use of its middle toe, which is curiously furnished with a serrated claw. (p. 78)

This passage is characteristic of White's prose in several key respects. First, and most importantly, it is concerned with nonhuman nature in action. A true understanding of the nightjar comes not from studying it as a decontextualised specimen, but from watching its movements within a particular habitat and in relation to other creatures. As intimated earlier in the chapter, there is a parallel here with Karen Barad's concept of 'intra-action', which imagines objects and agents as emerging through mutual relations rather than acting as pre-existing entities.[40] We are not suggesting, of course, that White and Barad are on the same page philosophically, but they do share an ontology that is fundamentally dynamic and relational. The second characteristic aspect of the passage is its use of terminology. White knows the bird's and the insect's Linnaean names – thus showing that he speaks the lingua franca of European naturalists – but also provides the vernacular designations by which they are known in a

[39] Dadswell, *The Selborne Pioneer*, p. 55. [40] Barad, *Meeting the Universe Halfway.*

particular place. Third, the passage is based around personal observation rather than laboratory work, and it emphasises the pleasure of field study. Fourth, White seeks to extrapolate general principles from his specific observation: in this example, an understanding of the purpose of the nightjar's 'serrated claw'.[41]

A further reason for White's interest in nightjars was that they are migrants only found in Britain during the summer. White was fascinated and concerned by bird migration – a particularly powerful example of nature in motion – throughout his life. In particular, he suspected that some hirundines did not migrate south at the end of the summer, but stayed in Britain (p. 174). He was therefore alert to stories of such birds being discovered in the winter (pp. 52–3) and speculated that they hibernated 'till the return of summer and fine weather awakens them' (p. 116); like many other eighteenth-century writers, he was even open to the idea that they 'conglobulated' (to use Samuel Johnson's term) under water. Anne K. Mellor shrewdly argues that, for White, 'the swallow was the essence of an idealised English national identity: domestic, devoted to family values, hard-working, loyal to the community, a powerful protector against foreign invasions, and aesthetically pleasing. This was why he could not bear to think of them annually abandoning his precious English countryside.'[42] This is plausible, but White saw the swallow as more than just a symbol of Englishness. As Bewell points out, 'the creatures that he values the most and that receive his greatest attention are those that are most fully integrated into his own social world'.[43] Swallows, swifts and house martins were particularly well integrated from White's perspective because of their tendency to nest in human dwellings. One of his greatest achievements was to understand humans and nonhumans as forming an intra-active community. Like all communities, this one had its limits and in White's case depended on an ideal of Englishness that found it difficult to accommodate apparent outsiders, such as Asian cockroaches or, more troublingly, gypsies (pp. 158–9). There is a nativist strand in British nature writing that has led to recent controversies and attempts to make the genre more diverse (see Chapter 4; also Afterword). The creative tension between British nature as static, hermetic, unchanging and nature as dynamic, porous and unpredictable is already apparent in White's work.

41 He was probably wrong about this; the claw is now thought to be used for preening rather than feeding.

42 Anne K. Mellor, 'The Baffling Swallow: Gilbert White, Charlotte Smith, and the Limits of Natural History', *Nineteenth-Century Contexts*, 31 (2009), 299–309 (p. 305).

43 Bewell, *Natures in Translation*, p. 176.

The dynamism of White's view of nature is reflected at the level of form as well as content. It is sometimes forgotten that his book was first published as *The Natural History and Antiquities of Selborne.* The natural history part comprises two sets of letters – to Thomas Pennant and Daines Barrington – written over a twenty-year period and makes up about two thirds of the original book. The second section of the book, on antiquities, comprises a single set of letters with no addressee, and is normally omitted in modern editions. The epistolary form is important, suggesting that White, although not exactly at the centre of things, conceived himself as a part of a developing conversation within a collaborative community of scholars in contrast to Rousseau's more individualistic understanding of natural history. The letters were carefully curated; as Anne Secord notes, 'some had been addressed to other correspondents, others were based on the translation of entries from White's journal into letter form, and even those that had originally been sent to the two naturalists were divided up into several shorter ones' (p. xx). White's journals also provided a vital urtext for the *Natural History.* Between 1768 and 1793, he wrote almost daily within the blank diaries designed by Barrington and published by Gilbert's brother Benjamin. The pages of the journal are split into nine columns for recording different kinds of natural-historical data. Of particular interest, though, are White's discursive notes, which frequently overrun the journal's systematic organisation.

Rhian Williams has argued that reading White's journals, as opposed to the curated letters that comprise *Selborne*, can help us to see his everyday parochialism as an 'ecologically significant mode'.[44] For Williams, the journal represents a coming together of individual field notes and more communal and systematic observation, and is therefore 'responsive to changing patterns and circumstances, grounded in familiar, local knowledge and lived experience' and 'brings many discourses together rather than privileging one perspective'.[45] Of course White had his prejudices and biases. But his willingness to inhabit different perspectives and modes of writing is noteworthy. *Selborne* even includes an inset poem, 'The Naturalist's Summer-Evening Walk', which provides a greater opportunity for Rousseauvian reflection than White's more scientific letters. As well as employing the 'curious mind', White notes, 'NATURE'S works … / Inspire a soothing melancholy joy' (p. 56). Brycchan Carey has recently

[44] Rhian Williams, 'Gilbert White's Eighteenth-Century Nature Journals As "Everyday" Ecology', *Interdisciplinary Studies in Literature and Environment*, 24 (2017), pp. 432–56 (p. 435).

[45] *Ibid.*, p. 452.

analysed how White utilises a range of literary techniques in his prose and poetry: allusion (particularly to Virgil and Milton), scansion, epistolary fiction, poetic diction, alliteration and vivid imagery. Even the brief entries in the journals at times manifest a 'curiously intense and poetic language' akin to the vivid compressions of imagist verse.[46]

White's writing derives energy not only through the incorporation of different kinds of writing, but also from the mobility of the narrative perspective, which switches deftly between different scales. For example, he writes about the response of crickets when he 'puts them into the crevices of a dry stone wall':

> For though they seem distressed by being taken out of their knowledge, yet the first that got possession of the chinks would seize upon any that were obtruded upon them with a vast row of serrated fangs. With their strong jaws, toothed like the shears of a lobster's claws, they perforate and round their regular holes, having no fore-claws to dig, like the mole-cricket. When taken in hand, I could not but wonder that they never offered to defend themselves, though armed with such formidable weapons. (p. 199)

One of the key similarities between White and subsequent nature writers is a willingness to find excitement and significance in apparently trivial creatures and occurrences. In this passage, the field cricket, initially diminished and 'distressed' by being removed from its familiar surroundings, is magnified into a kind of monstrosity rather like Robert Hooke's famous drawing of a flea in *Micrographia* (1665). Its fangs become 'vast', 'strong' and 'formidable weapons', even when in the palm of the naturalist's hand. The narrative perspective is mobile: zooming in on the fangs, then out to the naturalist, and then back in again. In another letter, White describes his experience of a 'wonderful shower' of cobwebs in a field. He notes that, contra superstition, this was 'the real production of small spiders, which swarm in the fields in fine weather in autumn, and have a power of shooting out webs from their tails so as to render themselves buoyant, and lighter than air' (pp. 155–6). Writing more than a century before Charles Darwin's celebrated monograph on the subject, White was also the first natural historian to recognise the vital ecological significance of earthworms, an apparently 'small and despicable link in the chain of nature' that would 'make a lamentable chasm' if lost (p. 172).

The final few letters of *Selborne* zoom out rather than in, to address the relationship between White's home and transnational phenomena by

[46] Carey, 'The Literary Gilbert White'.

describing the extreme weather that has affected the parish over the years. His account of a severe frost in 1768 causes him to reflect on the fate of imported plants and on the value of snow as a 'kindly mantle' for 'infant vegetation' in northern countries like Sweden (p. 225). In 1776, Selborne experienced 'rugged Siberian weather' through which 'the snow was driven into the most romantic and grotesque shapes, so striking to the imagination so as not to be seen without wonder and pleasure' (p. 226). A journey to London is more troubling. After travelling through 'a sort of Laplandian-scene, very wild and grotesque indeed', White discovers 'a still more singular appearance' in the metropolis. The snow muffles the sounds of carriages and horses, leading to a terrifying silence that conveys 'an uncomfortable idea of desolation' (p. 227). The potentially invasive and destabilising power of nonhuman nature is evoked most strongly in the book's penultimate letter:

> The summer of the year 1783 was an amazing and portentous one, and full of horrible phaenomena; for, besides the alarming meteors and tremendous thunder-storms that affrighted and distressed the different countries of this kingdom, the peculiar haze, or smokey fog, that prevailed for many weeks in this island, and in every part of Europe, and even beyond its limits, was a most extraordinary appearance, unlike any thing known within the memory of man … The sun, at noon, looked as blank as a clouded moon, and shed a rust-coloured ferruginous light on the ground, and floors of rooms; but was particularly lurid and blood-coloured at rising and setting … The country people began to look with superstitious awe at the red, louring aspect of the sun; and indeed there was reason for the most enlightened person to be apprehensive; for, all the while, Calabria and part of the isle of Sicily, were torn and convulsed by earthquakes; and about that juncture a volcano sprung out of the sea on the coast of Norway. (pp. 232–3)

Although White did not know it, the 'peculiar haze' was caused by a lengthy eruption of volcanic fissures around Mount Laki in Iceland that began in June 1783.[47] The lava did not directly affect populated areas, but led to famine and fluoride poisoning that is thought to have killed 10,000 people – about 20 per cent of Iceland's population at the time. The resulting toxic smog spread throughout Europe in the summer of 1783, causing respiratory problems and thousands of deaths. That summer was unusually hot and was also marked by other strange phenomena such as a major earthquake in Italy (killing 30,000 people) and meteor sightings,

[47] For an overview of the eruption and its effects, see Clive Oppenheimer, *Eruptions That Shook the World* (Cambridge: Cambridge University Press, 2011), chapter 12.

followed by a severely cold winter that the eruption probably caused.[48] It is hardly surprising that some people looked for religious explanations, fearing the Day of Judgement was at hand. Although White clearly sees himself as one of the 'enlightened' rather than one of the 'superstitious', he also implies that there is something ominous, perhaps even apocalyptic, in the coincidence of these 'horrible phaenomena'. Through allusions to apocalyptic passages from the Bible and *Paradise Lost*, White invokes a sense of spiritual and political as well as meteorological crisis. This is hardly surprising given that in the summer of 1783, Britain was reeling from its defeat in the American Revolutionary War. Indeed, another exemplary writer of English localism, William Cowper, would interpret the toxic atmosphere of 1783 as a sign of God's displeasure with the British Empire.[49] White does not go so far, but in its final pages *The Natural History of Selborne*, so often associated with a vision of England as a static Eden, invokes crisis and change.

Selborne's final letter continues the topic of extreme weather – a severe storm in June 1784 – but focuses on the local area: 'at about a quarter after two the storm began in the parish of Hartley, moving slowly from north to south; and from thence it came over Norton-farm, and so to Grange-farm, both in this parish' (p. 234). After the grand drama of the summer of 1783, the level of detail here may seem bathetic. But, as in the example of the field cricket, it shows White's ability to shift scalar perspective and see what is important in the apparently trivial. For it turns out that the two farms were 'greatly injured' by the storm, which began with 'prodigious torrents of rain', which were 'soon succeeded by round hail, and then by convex pieces of ice, which measured three inches in girth' (p. 234). Precise information about the storm is also important because its small size, 'about two miles in length and one in breadth', means that much of the neighbourhood is relatively unaffected. White notes that as he was sitting down to dinner, the storm reached its outer limit of the middle of Selborne: '[T]he hail broke my north windows, and all my garden-lights and hand-glasses, and many of my neighbours' windows. [We] were soon diverted from our repast by the clattering of tiles and the jingling of glass.' Nonhuman nature is not always domesticated in White's book; here it disrupts everyday life and

48 Laki may also have contributed to the very hot summer in Europe in 1783, although its effects on the global climate are still debated: see Alexandra Witze and Jeff Kanipe, *Island on Fire* (London: Profile, 2014), chapter 6.

49 For a more detailed account of literary responses to Laki, see David Higgins, 'British Romanticism and the Global Climate', in Adeline Johns-Putra (ed.), *Climate and Literature* (Cambridge: Cambridge University Press, 2019), pp. 128–43.

human constructions, leaving glass smashed and 'the hollow lane toward Alton' so 'torn and disordered as not to be passable' (p. 234). This concern with the catastrophic, inhuman power of the elements is apparent in later Romantic writing, particularly that surrounding the Tambora eruption of 1815 and the subsequent 'Year without a Summer'.[50] But whereas poets like Lord Byron and Percy Bysshe Shelley tended to be interested in the philosophical and existential aspects of such power, prose nature writers such as White focused more on its effects on everyday life.

Dorothy Wordsworth's journals offer a kind of midpoint between Rousseau's confession and White's natural history: more outward-looking than Rousseau, and more personal than White. Critics now understand Dorothy not as a mere adjunct to her brother William, but as a collaborator with him and a major author in her own right.[51] Neither of the Wordsworths was a natural historian to the extent of authors like White, Smith and Clare. However, they share with them an emphasis on nature as an active presence, as shown in this passage from Dorothy's Grasmere journal:

> We amused ourselves for a long time in watching the Breezes some as if they came from the bottom of the lake spread in a circle, brushing along the surface of the water, & growing more delicate, as it were thinner & of a *paler* colour till they died away—others spread out like a peacocks tail, & some went right forward this way & that in all directions. The lake was still where these breezes were not, but they made it all alive. I found a strawberry blossom in a rock, the little slender flowers had more courage than the green leaves, for *they* were but half expanded & half grown, but the blossom was fully spread out. I uprooted it rashly, & felt as if I had been committing an outrage, so I planted it again—it will have but a stormy life of it, but let it live if it can.[52]

The 'Breezes' here are not personified, nor are they represented by the kind of conventional neoclassical apparatus William railed against in the preface to *Lyrical Ballads* (1800). However, they are active forces, given body by their unpredictable effects on the surface of the water, but understood to exist separately from those effects. Thus, Dorothy imagines from the ripples that some breezes themselves are 'thinner' and '*paler*' than others.

50 See David Higgins, *British Romanticism, Climate Change, and the Anthropocene: Writing Tambora* (London: Palgrave, 2017).

51 See Lucy Newlyn, *William and Dorothy Wordsworth: 'All in Each Other'* (Oxford: Oxford University Press, 2013).

52 Dorothy Wordsworth, *The Grasmere and Alfoxden Journals*, ed. Pamela Woof (Oxford: Oxford University Press, 2002), p. 61. Further references are included in the text.

The emphasis on the second term suggests her awareness that this is a kind of poetic licence; breezes are not normally imagined as having colour but nonetheless appear embodied via the lake. Crucial to this passage is close, personal observation of apparently insignificant phenomena: apparent in the move from the breezes rippling on the water to the discovery of the 'little slender flowers' of the 'strawberry blossom'. Also noteworthy is the ambivalent role of the self: capable of destructive interference ('I uprooted it rashly'), but also of a more benign wise passiveness ('let it live if it can'). Nature writing generates much of its force through this creative tension.

The energy of nonhuman entities is emphasised throughout the journal. In the prose counterpart of William's more famous poem, Dorothy encounters 'daffodils so beautiful they grew among the mossy stones about & about them, some rested their heads upon these stones as on a pillow for weariness & the rest tossed & reeled & danced & seemed as if they verily laughed with the wind that blew upon them over the Lake, they looked so gay ever glancing ever changing' (p. 85). There is some personification and anthropomorphism here, but it is delicately done and does not detract from the specificity of the image. Dorothy typically describes an active, bustling landscape 'full of life & motion. The wind blew briskly & the lake was covered all over with Bright silver waves that were there each the twinkling of an eye, then others rose up & took their place as fast as they went away. The Rocks glittered in the sunshine, the crows & the Ravens were busy, & the thrushes & little Birds sang' (p. 80). This emphasis on energy, process and interconnection is the antithesis of the static Nature observed by the taxonomist, and is apparent in the account of a birch tree which

> was yielding to the gusty wind with all its tender twigs, the sun shone upon it & it glanced in the wind like a flying sunshiny shower—it was a tree in shape with stem & branches but it was like a Spirit of water—The sun went in & it resumed its purplish appearance the twigs still yielding to the wind but not so visibly to us. The other Birch trees that were near it looked bright & cheerful—but it was a Creature by its own self among them. (p. 40)

The association of individual natural forms with a spirit or genius is neoclassical. But the language here is dynamic rather than clichéd. Paradoxically, the tree's yielding to the wind is what makes it an active force. It is much more a process – 'a flying sunshiny shower' emerging through elemental intra-actions – than it is an object. To describe it as 'a Creature' suggests agency and creative capacity, with the slight implication that it figures the creativity of the writer herself, but the self is not imposed on to the landscape in Rousseauvian fashion.

Despite her emphasis on natural energies, the environments of Dorothy's journals are not wild places, but cultural landscapes in which humans and nonhumans intermingle:

> There was the gentle flowing of the stream, the glittering lively lake, green fields without a living creature to be seen on them, behind us, a flat pasture with 42 cattle feeding, to our left the road leading to the hamlet, no smoke there, the sun shone on the bare roofs. The people were at work ploughing, harrowing & sowing—Lasses spreading dung, a dogs barking now & then, cocks crowing, birds twittering, the snow in patches at the top of the highest hills, yellow palms, purple & green twigs on the Birches, ashes with their glittering spikes quite bare. The hawthorn a bright green with black stems under, the oak & the moss of the oak glossy. (p. 87)

This scene is replete with human and nonhuman activity, including even the brightness and glossiness of the hawthorn, oak and moss. In this georgic setting, agricultural labour seems to exist in harmony with its environment. Dorothy's journals invoke a sense of local community to a far greater degree than does White. Like some later women nature writers, such as Kathleen Jamie, Dorothy intermingles her close observations of nature with descriptions of domestic life and social encounters (see Chapter 4). Despite its secluded beauty, Grasmere was not unaffected by the conflicts of the Romantic period. The Wordsworths encountered wandering beggars, victims of economic distress, many of whom had been soldiers or sailors involved in the activities of empire. For example, in March 1802, an ex-mariner called Isaac Chapel knocked on their door and talked to them for two hours about the horrors of life on a 'Slave Ship' and the time he had spent among 'Indians' in North America (pp. 78–9). The Wordsworths themselves had a complex relationship to empire. Before moving to Grasmere, they spent a formative period of around two years living together rent-free at Racedown in Dorset, a property owned by John Praetor Pinney, a Bristol merchant and owner of plantations and slaves. They were friends with the leading abolitionist Thomas Clarkson and his wife Catherine, and their famous encounter with the daffodils in Grasmere occurred as they walked home from a visit to their house. William wrote against the slave trade during his radical youth, although his position became more ambivalent as he got older to the extent that he wrote against the 'immediate abolition' of slavery in the 1830s, while claiming to abhor the practice. His epic poem *The Prelude* (1850) was influenced not only by his experience of English landscapes, but also by travel narratives; its 'spots of time' have been discussed as simultaneously refuges from and products of imperial modernisation.[53] A similar ambivalence can be found in his popular poem *The Excursion* (1814). If the poem's opening represents the collapse of the English rural home as a result of Britain's imperial adventures, then the final book connects the education of the potentially rebellious lower classes of 'merry England' to the improving effects of British imperialism throughout the world.[54] Throughout the nineteenth century and beyond, images of English nature would offer a powerful justification for British exceptionalism and imperial expansion.[55]

Experiments in Form: Charlotte Smith and Thomas Bewick

Modern prose nature writing is a porous and complex amalgam of genres, including autobiography, travel writing and natural history, but is none-theless usually identifiable as distinctive from the elements that comprise it. Its distinctiveness is much less apparent in the Romantic period, when it is not so much a genre as a mode of engagement – combining personal and natural-historical perspectives – that manifests across and within different kinds of texts. Charlotte Smith (1749–1806) and Thomas Bewick (1753–1828) did not produce the kind of extended non-fictional prose essays we now associate with modern nature writing. However, they are key figures in its history. Both were fascinated by the dynamism and mobility of the nonhuman world, and its metonymic connections with human life. And they experimented with form in order to explore these connections: Smith produced rich texts that integrated prose and poetry, while Bewick integrated visual and verbal images of nature and culture.

Representations of the more-than-human world – based on personal experience and on wide reading – are crucial to the full range of Smith's texts: her poems, her educational works and her fiction. Due to the remarkable work of scholars over the past three decades, Smith is now understood as a major figure in British Romanticism.[56] However, her

[53] Carl Thompson, *The Suffering Traveller and the Romantic Imagination* (Oxford: Oxford University Press, 2007), chapter 5; Saree Makdisi, *Romantic Imperialism: Universal Empire and the Culture of Modernity* (Cambridge: Cambridge University Press, 1998).

[54] For a valuable account of empire in *The Excursion*, see Alison Hickey, 'Dark Characters, Native Grounds: Wordsworth's Imagination of Imperialism', in Alan Richardson and Sonia Hofkosh (eds.), *Romanticism, Race, and Imperial Culture, 1780–1834* (Bloomington: Indiana University Press, 1996), pp. 283–310.

[55] Ian Baucom, *Out of Place: Englishness, Empire, and the Locations of Identity* (Princeton, NJ: Princeton University Press, 1999).

[56] Jacqueline M. Labbe's work has been of particular importance; see, for example, *Charlotte Smith: Romanticism, Poetry and the Culture of Gender* (Manchester: Manchester University Press, 2003).

significance within modern nature writing is still not fully recognised. She deserves better, not least because of the distinctive natural-cultural discourse created by her experiments in form. She is particularly fond of two techniques: the inclusion of nature poems in the body of prose nature writing as a way of extending its moral and imaginative reach, and the inclusion of expansive prose annotations alongside the poetry as a way of grounding and evidencing its claims. A significant aspect of Smith's nature writing is its critical engagement with White and Rousseau. Like White's, her work is based on careful observation and knowledge of her local patch: in her case, the area around Brighton and the South Downs. Indeed, her depictions of birds, animals and plants have a specificity not always apparent in writings by the canonical male Romantic poets. Like Rousseau, she understands immersion in nonhuman nature as offering a cure for the ills of modernity. Her *Elegiac Sonnets* (1784–1800) and the prefaces to its various editions are written in the confessional mode he did so much to popularise.[57]

Smith had a very difficult life. Due to her father's financial problems, she was forced to marry the wealthy spendthrift Benjamin Smith at the age of fifteen. They had twelve children, only six of whom survived Charlotte. The marriage was deeply unhappy due to Benjamin's violence and bad behaviour. When her father-in-law died, problems with his will meant that Smith and her children did not receive the money they were due. She wrote the first volume of *Elegiac Sonnets* while confined to prison due to her husband's debts; the proceeds secured their release. She left Benjamin in 1787 and spent the rest of her life writing to support herself and her children. This biographical sketch is important to understanding Smith as a nature writer. Working under considerable financial pressure, and publishing under her own name within a male-dominated field, Smith often dramatised her struggles within her works and presented nonhuman nature as offering a temporary respite from her suffering. It in no way lessens the reality of that suffering to suggest that her writings, including paratextual material such as prefaces, epigraphs and frontispieces, dramatised her predicament for commercial effect. In this, of course, she followed Rousseau; like him, she understood her individual anguish within a broader political framework. The 1797 preface to *Elegiac Sonnets*, for example, ends with a reference to the Anglo-French conflict: 'the extensive

[57] We are not suggesting that Rousseau is the only important figure in the late eighteenth-century discourse of sensibility. Goethe's sentimental novel *The Sorrows of Young Werther* (1774), for example, is a key intertext for Smith's sonnets: see Elizabeth A. Dolan, *Seeing Suffering in Women's Literature of the Romantic Era* (Aldershot: Ashgate, 2008), chapter 1.

and still threatening desolation, that overspreads this country, and in some degree, every quarter of the world'.[58]

In the face of private and public woes, Smith's autobiographical personae, like Rousseau's, take refuge in the more-than-human world and find botany a particularly effective form of *askesis*.[59] In Sonnet 79, entitled 'To the Goddess of Botany', the speaker desires to 'find shelter' from 'Folly ... Violence and Fraud' among the Goddess's 'silent shades of smoothing hue'. She particularly emphasises the huge variety of attractions offered by botany and the opportunity it provides to escape into 'wilds remote' (p. 82). An extensive footnote to the title – much longer than the sonnet itself – refers extensively to Rousseau's 'Seventh Walk' and the botanical nature cure. Smith notes that '*I* cannot now turn to any other pursuit that for a moment sooths my wounded mind', with the italicised first person emphasising that she is writing in confessional mode (p. 82). She quotes three passages from *Rêveries* in which Rousseau describes the capacity of botany to amuse, distract and smooth the troubled subject; how it forces him to stop himself ruminating on his misfortunes and to keep his imagination in check; and the power of 'brilliant flowers, diverse colours of meadows, fresh shady spots, brooks, thickets, [and] greenery' to 'purify' his imagination (p. 232). Although she is careful not to claim to be equal to Rousseau's 'talents' or his 'excessive irritability', she notes the similarity in their situations: 'it has been my misfortune to have endured real calamities that have disqualified me from finding any enjoyment in the pursuits and pleasures which occupy the generality of the world' (p. 82). Those 'calamities', however, would at times be too much even for botany to ameliorate. In Sonnet 65, the 'botanic pencil's mimic powers' fail as a result of the death of her daughter, Anna Augusta (p. 74). And although the speaker of Sonnet 91 can still 'in groups these mimic flowers compose' (p. 89), the closing sestet focuses on the image of her daughter that 'ceaseless grief' has 'enshrined' in 'thy Mother's heart' (p. 89). Smith's work meditates self-consciously on the therapeutic potential of ecomimesis, but there are moments when the writing of grief erases the writing of nature.[60]

[58] Stuart Curran (ed.), *The Works of Charlotte Smith*, 14 vols. (London: Pickering and Chatto, 2005–7), vol. 14, ed. Jacqueline M. Labbe, p. 68. Further references to Smith's poetry are to this volume and are in the text.

[59] For a detailed account of Smith's interest in the therapeutic potential of botany, see Dolan, *Seeing Suffering*, chapter 4.

[60] See also Smith's novel, *The Letters of a Solitary Wanderer*, the title of which clearly alludes to Rousseau's *Reveries*. Its narrator is so tormented that he finds that botany 'used to interest and soothe my mind beyond any other study; but now I am in a state of spirits where it would depress

Although women were involved in all aspects of botany in the Romantic period, they had to negotiate both the long-standing 'iconographic tradition that rendered women as passive flowers' and the conservative suspicion that such studies – particularly when based on the Linnaean system with its emphasis on sexual parts – had a corrupting effect on femininity.[61] Theresa M. Kelley notes the strategic 'decorousness and moral fervour' with which Smith wrote about botany, but also the playful and complex way in which she combined Linnaean taxonomy with more figurative and vernacular names that emphasised repetition rather than differentiation.[62] Botany is a key component in the natural-historical education outlined in the *Conversations Introducing Poetry: Chiefly on the Subjects of Natural History. For the Use of Children and Young Persons* (1804), a text which presents itself as both an anthology of poetry and a kind of collection of plants and animals.[63] Like *Selborne*, it emphasises nature in motion and is suspicious of more static taxonomy. Reflecting on a visit to the Leverian Museum in London, the book's main interlocutor, Mrs Talbot, notes that:

> The birds, or insects, or quadrupeds, though they may be very well preserved, lose that spirit and brilliance, which living objects only can possess. The attitudes of the birds are stiff and forced, and without their natural accompaniments ... Their formal or awkward appearances, when stuffed or set on wires, always convey to my mind ideas of the sufferings of the poor birds when they were caught and killed, and the disagreeable operations of embowelling and drying them.[64]

Mrs Talbot identifies two related problems with displays of taxidermy. One is ethical: taxidermy reminds one of the death and 'sufferings' of the creatures portrayed, as well as the indignities inflicted on their bodies.[65] This view is in keeping with the text's other criticisms of animal cruelty, embodied by the thoughtless and spoilt rich child Harry Scamperville (pp. 72–4). The other problem is ecomimetic. Static taxidermic

rather than charm them': Curran (ed.), *The Works of Charlotte Smith*, vol. 11, ed. David Lorne Macdonald, p. 6.

61 Theresa M. Kelley, *Clandestine Marriage: Botany and Romantic Culture* (Baltimore, MD: Johns Hopkins University Press, 2012), chapter 4 (p. 93).

62 *Ibid.*, pp. 118–25.

63 Dolan, *Seeing Suffering*, p. 120. Botany is also important to Smith's other educational works, particularly *Minor Morals* (1798), as well as her novel *The Young Philosopher* (1798).

64 Curran (ed.), *The Works of Charlotte Smith*, vol. 13, ed. Judith Pascoe, p. 179. Further references to the *Conversations* are included in the text.

65 For an astute discussion, see Richard De Ritter, '"Wonderful Objects" and "Disagreeable Operations": Encountering the Leverian Museum in Writing for Children', *Journal for Eighteenth-Century Studies*, 41 (2018), 427–45.

constructions of animals and birds outside their habitats, even if visually arresting, do a poor job of capturing what is important about them. The implication is that their 'spirit and brilliance' can be better represented through writing.

However, Mrs Talbot takes a different view of collections of plants, which offer 'only pleasing ideas', for both aesthetic and practical reasons. She implicitly takes issue with Rousseau, noting that 'I cannot say that I think the pleasure of botanizing destroyed by considering plants as convertible into drugs', and emphasises how the objects around us are manufactured with 'the produce of vegetables' (pp. 179–80). Thus the beauty and utility of plants go hand in hand, whereas they are potentially in tension when it comes to birds and animals.

The *Conversations* ends with 'the last Poem we have got to complete our collection' (p. 230): a mock-heroic account of 'Flora' that is self-consciously set up by the preceding prose dialogue. It begins with a discussion between Mrs Talbot and Emily about the importance of 'the heathen mythology' in developing poetical taste and the dangers of reading Ovid ('by no means the most proper among the Roman poets'). Mrs Talbot takes Emily briefly through epic before discussing the 'mock heroic' of Pope and Hayley (pp. 223–5). She notes that their 'machinery' of 'Sylphs and Gnomes' is also used in Erasmus Darwin's 'magnificent poem' *The Botanic Garden* (1791) (p. 225); this praise is noteworthy given that conservative commentators saw the poem as politically and sexually suspect. The discussion of poetry is a prelude to 'our little cabinet picture of Flora' (p. 230) that George and Mrs Talbot have painted; while the machinery may be fanciful, Mrs Talbot emphasises that the poem is based on scholarship and fieldwork. Like the other poems in the book, 'Flora' is imagined as part of a botanical 'collection', but it is also itself a collection of plants in which scientific and 'common' names are utilised as 'the structure of the verse' demands. It begins with a Rousseauvian ideal of retreat from 'the crimes and follies' of mankind and celebrates the power of Fancy's 'magic pencil' to recover 'lovely forms' from memory (pp. 230–7). The poem therefore mimics the function of Rousseau's herbaria.

The *Conversations* shows Smith's knowledge of the technical natural history of the eighteenth century, including the works of Linnaeus and his popularisers. But her work is also usefully understood in relation to the more localised, unsystematic and personal tradition pioneered by White and Rousseau. For example, the sixth conversation begins with Mrs Talbot and her charges journeying on horseback (p. 153). This leads to a reflection on the expense of keeping horses and the decline of the middle-class standard of living due to 'the accumulation of wealth, which has made a great deal of difference in the value of money; and wars, which have cost such immense sums to the nation' (p. 154). In such straitened circumstances, Mrs Talbot notes, it is particularly important to develop 'a taste for those pure and innocent enjoyments that are always to be found' and teach us – quoting Alexander Pope and emphasising Smith's adherence to the natural-theological tradition – 'To look thro' Nature up to Nature's God' (p. 154). It is early morning, and George notices 'those fine cobwebs covered with dew-drops, like so many globes of silver – how the net work is spun from one of those furze bushes to the other!'. Mrs Talbot informs them – and there is a footnote to White – that 'webs of the same fine structure … carry those minute spiders into the air, and that also weave the substance which sometimes is said to fall in showers … and is called gossamer' (p. 154). She then quotes Smith's sonnet 'To the Insect of the Gossamer'. Like some of her other poems, it begins with a precise description of a natural phenomenon: 'Small, viewless Aeronaut, that by the line / Of Gossamer suspended, in mid air, / Float'st on a sun-beam'. It then asks questions that imply the limits of human understanding of nonhuman 'design[s]'; a key theme in Smith's work. The sonnet turns in the sestet by finding an analogy between the spider's vulnerability and ephemerality, and the 'radiant dreams' of 'the young and visionary poet' (p. 155). This might be read as a kind of anthropocentric shift in which describing the insect of the gossamer turns out to shed light on human experience (metaphor). However, given that the bulk of the poem concerns the spider, the final lines are more usefully interpreted as emphasising the connection between human and nonhuman experience (metonymy).

Indeed, this connection is highlighted when Emily asks why the spider is referred to as an 'aeronaut' in the sonnet's first line. Mrs Talbot points out that the word does not mean simply a balloonist, but 'those adventurers who have learned to float *in the air*, by *means of the air*' (p. 155). It may therefore be applied to a spider as well as a person. Aptly, the poem is recited as they ascend a hill and the discussion of it concludes as they reach the top and contemplate the English Channel. Ballooning had been pioneered by the French in the early 1780s, with the first Channel crossing achieved in 1785. The term 'aeronaut' had been given a pointedly political charge by Edmund Burke in 1790, when he imagined the French Revolution as the 'desperate flights' of 'the aëronauts of France'.[66] Smith's sonnet is not in itself a commentary on politics, but it exists within a

[66] Edmund Burke, *Reflections on the Revolution in France* (London: J. Dodsley, 1790), p. 355.

geopolitical context that the format of the *Conversations* allows her to acknowledge. As in later nature writing, travel through a landscape provides a metonymic structure in which physical movement enables discursive shifts and imaginative leaps. From the top of the hill, Smith's characters see some ships far away, and Mrs Talbot alludes to the contemporaneous Anglo-French conflict by remarking on the dangers of being attacked by the enemy as they go down the Channel. She reflects on British commercial superiority over other nations, but also on the toll this takes on the families (including Smith's) of those who go to sea and to 'garrison' colonies. Then there is another shift; prompted by the sight of the 'small white sails', she reminds George of a poem on 'The Nautilus'. The poem is quoted after a brief prose description of the species and imagines nautiluses as 'mimic ships' due to their unique means of propulsion: 'The Nautilus expands his sail, / And scuds before the freshn'ing breeze' (p. 157). If 'tempests rend the sky' then he will retreat to the 'ocean's shelly bed', which offers an edifying analogy for the value of solitary repose in human life.[67]

This relatively short passage passes through a rich set of natural-cultural connections. Travelling on horseback leads to reflections on the state of the nation and natural theology. Encountering the gossamer recalls White's observations and Smith's poem. The spider is like a human aeronaut and the prospect that Mrs Talbot and her charges see is like what a balloonist would view. The ships remind them of the suffering caused by conflict, but also of the peculiar natural history of the nautilus: a 'mimic ship' that carries an edifying moral message. Ships, balloons, aeronautical spiders, scudding nautiluses and Smith's writing itself all share a peculiar and arresting buoyancy. The didacticism of the *Conversations* marks it out as a late eighteenth-century educational text, but its fluent shifts between nature and culture – enabled by the physical and imaginative mobility of its characters – are akin to what we would expect from modern nature writing.

Smith takes a similar approach in the posthumously published *A Natural History of Birds, Intended Chiefly for Young Persons* (1807). Although it is organised according to the Linnaean system, it goes beyond taxonomic description to encompass personal observations, extensive poetic quotations, Smith's own poems and retellings of classical myths involving birds. It shows natural and cultural history as intertwined.

67 The species to which Smith is referring, *Argonauta Argo*, is sometimes referred to as a 'paper nautilus', but is actually a pelagic octopus rather than a nautilus. It does not propel itself in the way Smith identifies, but the egg cases its females produce do provide buoyancy.

For example, her account of the raven begins with a brief description of its nesting and feeding habits, quotes Shakespeare's reference to its portentous call in *Macbeth*, and then balances this with White's description of the raven's parental affections. Smith notes that she finds 'a *species* of enjoyment' in listening to 'the hoarse voices of ravens': 'their deep and solemn cawing' seems to evoke a kind of ancient wisdom and continuity to the past 'as they are said to live to a great age'. She quotes a modern fable of the raven by Cowper before briefly mentioning its presence in Greek myth. Like White and Clare, Smith is particularly intrigued by the nightjar. Her discussion takes in its variety of names, its feeding on the wing, its use in popular superstition and her own observations of its appearance and habits. She notes that White mentions it in his poem 'The Naturalist's Walk', and also quotes four lines from her own sonnet in which it features. The bird also appears in her remarkable prospect poem 'Beachy Head' (1807), through a brief reference to the 'evening sounds' of the 'night-jar, chasing fern-flies' (p. 170). In a lengthy footnote, Smith notes that 'it is called the Night Hawk, the Jar Bird, the Churn Owl, and the Fern Owl, from its feeding on the Scaraboeus solstitialis, or Fern Chafer, which it catches on the wing with its claws, the middle toe of which is long and curiously serrated, on purpose to hold them'; she is here likely drawing on White's (not entirely accurate) account. She cites her own sonnet as the only other poem in which the bird's 'remarkable' sound is mentioned, and she goes on to correct herself for having suggested in that poem that it could be encountered in November. Partly repeating material from the *Natural History of Birds*, she notes its appearance in folklore and the 'abhorrence' in which it is held by 'peasants', who believe that it hurts livestock. There is no doubt some strategic self-plagiarism in the appearance of the nightjar across several of Smith's texts, but this also suggests something about the mobility of the bird itself.

The *Natural History* moves fluidly between prose and poetry, closing with the first appearance in print of Smith's poem 'To the Swallow': an ambivalent intervention in the debate about hirundine migration which had so fascinated White and which 'remained unresolved at the end of the eighteenth century'.[68] Smith followed White in being attracted to the swallow as an exemplar of native domesticity but, as Elizabeth Heckendorn Cook suggests, she was also intrigued by its mobility and exoticism.[69] The

[68] Elizabeth Heckendorn Cook, 'Charlotte Smith and "The Swallow": Migration and Romantic Authorship', *Huntingdon Library Quarterly*, 72 (2009), 48–67 (p. 60).
[69] *Ibid.*, p. 63.

poem begins with a fairly conventional description of the swallow's arrival. Its domesticity is emphasised through an invitation for it to 'attach / To my reed roof your nest of clay'. But the poem becomes more complicated as the speaker speculates on what could be learned if it was possible to 'translate' birdsong 'as if 'twere mark'd in written page'. She asks if the bird would report on its migration to exotic places or whether it would tell of staying in England to hide in a 'cliff's excaved brow' or under water, and concludes by suggesting that these questions are ultimately undecidable:[70]

Alas! how little can be known,
Her sacred veil where Nature draws;
Let baffled Science humbly own,
Her mysteries understood alone,
By *Him* who gives her laws.

'The Swallow' exemplifies Smith's willingness to assert her intellectual authority throughout her nature writing. But the end of the poem, which is also the ending of the *Natural History of Birds*, also addresses the limits of human authority and of the 'vain systems', including natural history, by which we try to make sense of the world. The appearance of the 'Indian sage' with the fabulous power of transcribing and interpreting 'the wild birds' song' offers an ironic undercutting of the project of the nature writer, whose words are bound to fall short even of the approximations provided by translation.

Thomas Bewick (1753–1828) was of the same generation as Smith, and like her experimented with form in an attempt to depict living nature.[71] In his most important work, the two-volume *British Birds* (1797–1805), he created an encyclopaedia that combined lively visual and verbal representations of birds.[72] His mastery of wood-engraving techniques allowed him to produce images with remarkable detail and flow. The entry for each species begins with an engraved image, with its common name, other local variants – for example, 'Bog-Bumper, Bitter-Bum, or Mire-Drum' for the bittern – and its Latin names in Linnaeus and Buffon.[73] There usually

[70] For an excellent account of the poem's engagement with natural history, see Mellor, 'The Baffling Swallow'.

[71] For general studies, see Iain Bain, *The Watercolours and Drawings of Thomas Bewick and His Workshop Apprentices*, 2 vols. (London: Gordon Fraser, 1981); Diana Donald, *The Art of Thomas Bewick* (London: Reaktion, 2013); Jenny Uglow, *Nature's Engraver: A Life of Thomas Bewick* (London: Faber, 2006). For a good online selection of images by Bewick, see the website of The Bewick Society: www.bewicksociety.org/Gallery.html (accessed 19 April 2021).

[72] The first volume was co-written with Ralph Beilby.

[73] Thomas Bewick, *A History of British Birds*, 2 vols. (London: The Folio Society, 2010), vol. 2, p. 22. Further references are included in the text.

follows a physical description of the bird, and then an account of its habitat and habits: what we would now call its ecology (see Chapter 3). As Katey Castellano notes, 'Bewick acknowledges systematic natural history, but he also remains faithful to a stubborn provincialism that highlights the boundaries of his own knowledge; for the most part, his engraved birds are embedded in particularised environments in or around Newcastle.'[74] Bewick was a keen observer of living creatures; however, given the range of the species depicted, and the rarity of some of them, he often had to rely on dead specimens. He much preferred recently shot birds to stuffed ones on account of 'no regard having been paid at that time to place the former in their proper attitudes, nor to place the different series of the feathers, so as to fall properly on each other. This has always given me a great deal of trouble.'[75] Bewick's birds, generally depicted in profile on a branch or a rock, are presented as very much alive. Although he acknowledged that 'every attempt to depicture nature must fall short of the original' (vol. 2, p. xiv), he also understood his work within a natural-theological tradition through which reading 'the great book of Nature' was morally edifying and brought one closer to God, for 'the hand of Providence [was] in every page' (vol. 2, p. xiii; see also Chapter 2). 'A good naturalist', he asserted, 'cannot be a bad man'; revealing an embryonic conservationist ethic, he hoped that readers would 'on viewing the portraits, spare the originals' (vol. 2, p. xiv). Even apparently destructive 'pests' such as sparrows, he argued, contribute to 'the grand economy of nature' and should be cherished rather than persecuted (vol. 1, p. 185).

Absolutely crucial to Bewick's aesthetics and his understanding of natural history is his inclusion of tailpieces – tiny, detailed vignettes of rural life – after nearly every entry in both volumes of *British Birds*. On a pragmatic level, they fill blank space and make the book more attractive. Crucially, they bring human beings into natural history, showing nature and culture as entangled even as they reverse anthropocentric polarities by miniaturising the human world and enlarging our view of birds.[76] The tailpieces do not usually directly connect to the species they follow, but rather contribute to the book's general rural ambience. They are not always decorous. As Diana Donald has argued, they perform many functions, including challenging the aesthetics of the picturesque, drawing attention to the failures of paternalistic governance, indicting human cruelty to

[74] Katey Castellano, *The Ecology of British Romantic Conservatism, 1790–1837* (London: Palgrave, 2013), p. 66.
[75] Thomas Bewick, *My Life*, ed. Iain Bain (London: Folio, 1981), p. 138.
[76] Castellano, *Ecology*, p. 79.

nonhumans and celebrating the independent spirit of rural people.[77] One depicts a man in travelling garb urinating against a wall, a process only fully revealed through his shadow. Like many of the other images in the book, the tailpiece is framed by trees and bushes, suggesting a sequestered rural nook, but not one that is idealised or hermetically sealed off from the wider world. In a similar nook, an old man is in the process of hanging a cat, perhaps because it has stolen food from him. The carnivalesque image following the entry for the pratincole depicts an older boy about to beat a dog that, along with a cat, has been harnessed to a small cart containing a younger boy who does not seem to be enjoying the situation. Another image shows a hunter with a gun strapped to his back using stilts to cross a river. The remarkable tailpiece to the redstart depicts a chimney sweep, dark all over, eating bread outside the gates of an estate. His black face connects him to the black-faced bird, but also acts as a reminder of the coal industry that was an important part of Bewick's landscape. Similarly, the vignettes on the title pages of both volumes of *British Birds* include a smoking colliery in the background.

Bewick saw birds as repositories of freedom at a time when rural people's customary practices and common access to land were under threat from the forces of modernisation.[78] His politics were complex, but he was highly critical of the mistreatment of the poor, especially through enclosure, and what he saw as a failure of higher ranks to live up to their paternalistic responsibilities. He also identified with the freedom of birds; in his posthumously published memoir, he evokes his 'active wild disposition' as a child: running naked across fells, fishing, climbing trees. But he also describes this growing talent for drawing the creatures around him: 'the beasts and birds which enlivened the beautiful scenery of woods and wilds surrounding my native hamlet, furnished me with an endless supply of subjects'.[79] Hunting was very much part of his cultural milieu, as well as casual forms of animal cruelty, but so was compassion and fellow feeling with 'intimate acquaintances' such as robins and sparrows and with 'rare visitants, such as the woodcock, the snipe and other waders'.[80] As with other nature writers, the mobility of birds was a source of considerable fascination for Bewick. Although he agreed that migration was a subject of 'great difficulty', he noted that birds 'glide through the air with a quickness superior to the swiftest quadruped' and therefore would be able to travel 'great distances' (vol. 1, pp. xl–xli). He was fascinated by the idea that they

77 Donald, *Art of Thomas Bewick*, chapter 2. 78 Castellano, *Ecology*, p. 66.
79 Bewick, *My Life*, p. 29. 80 *Ibid.*, pp. 29–38.

could even penetrate the 'Arctic Zone', 'forlorn regions of *unknowable* dreary space' that were beyond human powers of inquiry and could only be imagined (vol. 2, pp. xxiv). A key way in which nature writing differs from technical natural history is its emphasis on the limits of human epistemologies and the mysterious, withheld core of wild creatures.

The entry for one particularly mysterious bird, the 'Nightjar', reveals many of the qualities of Bewick's writing and art. The image shows the bird in profile, calling on a branch. It is an excellent subject for wood engraving because of its intricate grey, black and brown markings. Therefore little colour is sacrificed; the nightjar's camouflage connects it both to the tree branch on which it sits and the form in which it is represented. Bewick provides a fine depiction in which even small details such as the bristles on its upper mandible can be clearly made out. As usual, he provides its Latin name in Linnaeus, its French name in Buffon, and several common British names: 'GOAT-SUCKER, DOR-HAWK, OR FERN OWL' (vol. 1, p. 302).[81] However, he was keen that 'nightjar' be adopted: 'To avoid … perpetuating error, we have dropped the name Goat-sucker, which has no foundation but in ignorance and superstition' (vol. 1, p. 303). Although Bewick was sympathetic to rural customs and aware of the cultural significance of birds, he also had no qualms about correcting 'error' when he found it. He prefers nightjar because the name 'bears some analogy to the nature and qualities of the bird, both with respect to the time of its appearance, which is always the dusk of the evening, as well as to the jarring noise which it utters' (vol. 1, p. 303).

Onomatopoeic names show language not only as representing the nonhuman, but connected to it through mimicry. This connection is also emphasised by another vernacular name – 'Wheel-bird' – that Bewick mentions later in the entry. The name ostensibly derives from its flight call: 'a sort of buzzing while on the wing, which has been compared to the noise caused by the quick rotation of a spinning-wheel'. It also reflects the flight pattern itself: 'their motions are irregular and rapid, sometimes wheeling in quick succession round a tree or other object, diving at intervals as if to catch their prey, and then rising again as suddenly' (vol. 1, p. 304). Like other Romantic nature writers, Bewick is interested in the nightjar not as an inert specimen but as an active being, with rapid and sudden changes of direction that have the power to surprise. Its 'wheeling' physical energy is reflected not just in its movements, but in

[81] The name dor-hawk comes from one of the nightjar's food sources: the cockchafer or dor-beetle. Bewick follows White in imagining how the bird catches such delicacies on the wing.

its 'jarring' call and one of its vernacular names. This name also associates the bird with a machine of human manufacture and of great importance to rural communities. Once again, human and natural history are entwined. It is fitting, therefore, that the tiny tailpiece depicts three items of human manufacture: a hat, a walking stick and a flagon or jar.

Bewick was a highly self-conscious artist and one of the pleasures of *British Birds* is its playfulness with form. As Castellano points out, 'many of Bewick's engravings depict other engravings, whether of mile markers, boundary stones, or memorials', reflecting his belief in the communal and 'social potential of engraving and inscription', including his own work.[82] There are also moments in which ecomimesis is deliberately broken through explicit artifice. Particularly noteworthy is the image in which a miniature rural scene is largely blotted by a thumbprint (Figure 1.1), which

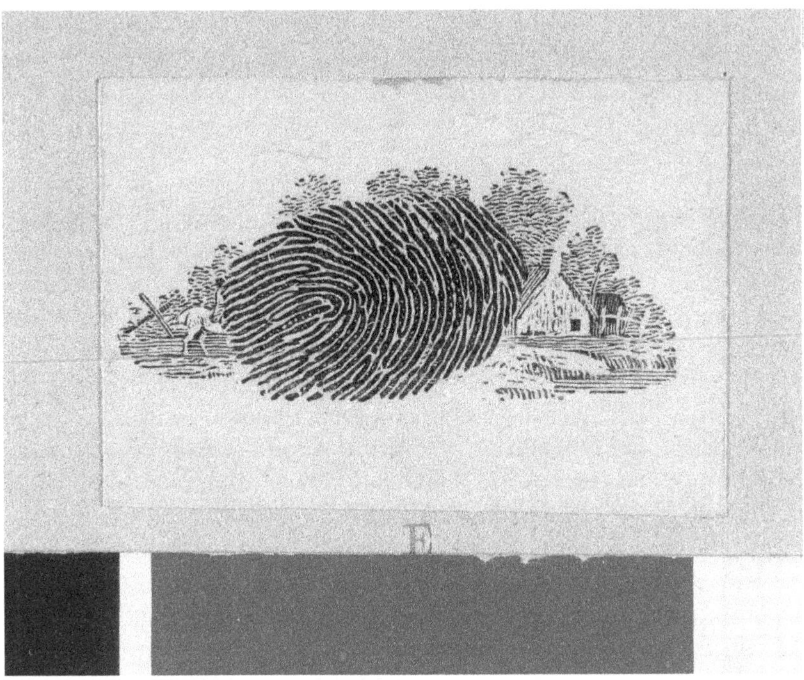

Figure 1.1 Thomas Bewick, tailpiece from *A History of British Birds* (Newcastle, 1797). © The Trustees of the British Museum

[82] Castellano, *Ecology*, p. 89.

Figure 1.2 Thomas Bewick, tailpiece from *A History of British Birds* (Newcastle, 1797). © The Trustees of the British Museum

'functions as a parabasis, a disruption of the artistic illusion'.[83] The tailpiece to the redwing, an autumn migrant, works similarly (Figure 1.2). It depicts a man and a woman sharing a horseback ride across a rural scene. But across a large part of the image, a leaf (probably from a hawthorn, one of the redwing's principal food sources) is superimposed, as if it has suddenly blown across the viewer's line of sight or (anachronistically) into her camera lens. Castellano suggests that both images draw attention 'to the shift in perspective required by observation of the miniature', by shrinking the landscape even further and 'highlighting the complicated and original designs that one can find in the smallest parts of life'.[84] This is true, but the images also draw attention to the ecomimetic limitations even of Bewick's art. They remind the viewer that she is at a considerable distance from the rural scenes and birds represented. They also emphasise that the images are carefully framed in not two, but three dimensions. By drawing attention to the front edge of the image, the leaf and the fingerprint create a kind of translucent barrier between image and viewer. Bewick's craft gave his

83 Anne K. Mellor, *English Romantic Irony* (Cambridge, MA: Harvard University Press, 1980), p. ix.
84 Castellano, *Ecology*, p. 78.

depictions of birds and animals a remarkable liveliness, but could not entirely disassociate itself from the objectifying gaze of the taxonomist.

John Clare, James Hogg and the Nature of Loss

In *The Country and the City* (1973), Raymond Williams identified nature writing's tendency to look backwards to a lost time of plenty, often identified with the author's childhood and the experience of the previous generation of rural dwellers. While attending to the very real histories of change and dispossession in the countryside, Williams pointed to a structural nostalgia in nature writing that he related to the interplay of pastoral and counter-pastoral.[85] One of Williams' key authors is John Clare, an author who celebrated and loved the more-than-human world, but who also articulated an agonising sense of loss conditioned in part by the enclosures affecting his local area, which started in 1809, when he was sixteen years old. Clare's writing is often elegiac and analeptic. Of course this is also characteristic of present-day nature writing, with the caveat that such writing is also increasingly preoccupied by future as well as past losses, offering a kind of proleptic elegy (see Chapter 4).[86] Local and global environmental changes are happening with a rapidity even Clare could not have imagined. When Williams looked at British nature writing of the past few centuries, he saw an 'escalator' in which texts of each generation would idealise the way of life of the previous generation. When we read past nature writing from the perspective of the Anthropocene, we also find ourselves on an escalator travelling from a relatively biodiverse past into a denuded future. Despite his preoccupation with loss, Clare describes a plenitude of local flora and fauna that would be impossible to find today in his home county of Northamptonshire, or indeed almost anywhere in the British Isles.

As Fiona Stafford suggests, both Clare and Hogg, as authors with backgrounds in agricultural labour, 'offer a lament for a world under threat from modernity but, in doing so, they develop exciting new literary styles which are perfectly suited to the demands of a modern, predominantly urban reading public'.[87] It is worth adding that they also had an ambivalent relationship with metropolitan literary culture through their associations

85 Raymond Williams, *The Country and the City* (London: Vintage, 2016 [1973]), chapters 2 and 3.

86 The phrase 'proleptic elegy' is from Patrick Brantlinger's *Dark Vanishings: Discourse on the Extinction of Primitive Races, 1800–1930* (Ithaca, NY: Cornell University Press, 2003).

87 Fiona Stafford, 'Pastoral Elegy in the 1820s: *The Shepherd's Calendar*', *Victoriographies*, 2 (2012), 103–27 (p. 104).

with journals and publishers that simultaneously feted and patronised them as 'untaught geniuses'. Clare has deservedly gained attention from critics and nature writers in recent years for his remarkable poetry. His prose, however, remains largely underappreciated. In many respects, it is apiece with his poetry, and indeed it was often where he tried out images later developed into poems. Both modes of writing reveal his remarkable capacity for observation, his interest in the apparently insignificant aspects of nonhuman nature, his deep love and sympathy for flora and fauna, his extensive knowledge of natural history, particularly of his local area, his concern with the environmental politics of landownership and access, and his self-consciousness about the challenges of nature writing and the limits of ecomimesis.[88] Clare's prose nature writing has not had the impact it deserves, in part because it comprises a set of unpublished fragments rather than a coherent book like White's *Selborne*, and was not published during his lifetime. These fragments were collected in Margaret Grainger's 1983 edition of *The Natural History Prose Writings of John Clare*. In what follows, we focus largely on the texts (especially fourteen draft letters), which Grainger suggests Clare intended for eventual publication as part of a natural history of his home village of Helpstone with his London-based publisher, James Hessey.[89] The following discussion particularly addresses Clare's concern with what made nature writing accurate and effective and his interest in how language could go beyond *representing* the more-than-human world by *connecting* us to it.

Clare admired authors whom he thought 'went to nature for their images' rather than relying on 'the old thread bare epithets' he associated with the tradition of 'Pastoral & descriptive poesy'.[90] Among those authors was Charlotte Smith; in his essay on the nightjar, he writes of having 'the pleasure to meet with [her sonnets] in a friends book case'. There is (gendered) condescension in his description of them as 'only pretty', but

88 For a pointed critique of the dangers of ecological approaches to Clare that are inattentive to his historical circumstances and his self-consciousness about his own creative practice, see Simon Kövesi, *John Clare: Nature, Criticism and History* (London: Palgrave, 2017). We have tried to avoid these errors.

89 The project was never completed; Robert Heyes has challenged Grainger's account, arguing that the letters were directed at different correspondents and were not connected to the proposed 'Natural History of Helpstone': see 'John Clare's *Natural History*', in Simon Kövesi and Scott McEathron (eds.), *New Essays on John Clare: Poetry, Culture, and Community* (Cambridge: Cambridge University Press, 2015), chapter 8.

90 John Clare, *The Natural History Writings of John Clare*, ed. Margaret Grainger (Oxford: Clarendon, 1983), pp. 39, 51. Clare rarely used punctuation. We have generally transcribed passages from this edition verbatim but, for the sake of readability, have occasionally omitted bracketed additions reflecting edits or confusions in the manuscript. Further references are included in the text.

he also notes that Smith 'wrote more from what she had seen of nature then from what she had read of it'. Therefore, her readers would find 'new images which they had not read before although they had often felt them'; for Clare, such poetry provides the 'happiest' pleasure (p. 34). The image of him 'meet[ing] with' Smith's poems suggests his fascination with the idea that nature writing could be constituted by presence rather than absence through generating a poetics of encounter.[91] It is therefore entirely fitting that the encounter with Smith's sonnets is directly followed by a description of an encounter in which 'he met with another nocturnal bird called a "night hawk"' (p. 34).[92] The fantasy – and Clare knows it is one – is of nature writing that is as transparent as an open window:

> The rainy morning has kept me at home & I have amused my self heartily sitting under Waltons Sycamore tree hearing him discourse of fish ponds & fishing what a delightful book it is the best English Pastoral that can be written the descriptions are nature unsullied by fashionable tastes of the time they are simply true & like the Pastoral ballads of Bloomfield breath of the common air & the grass & the sky one may almost hear the water of the river Lea ripple along & the grass & the flags grow & rustle in the pages that speak of it I have never read a happier Poem in my time. (p. 174)

The allusion is to Izaak Walton's *The Compleat Angler* (1653), a formally complex text that was a product of the turbulent political period after the English Civil War. Walton's discoursing anglers shelter under a sycamore during a rain shower; the power of good nature writing is to allow the reader to share such experiences. The truest pastoral, for Clare, is not an idealised picture of an imagined past, but an evocation of what is best and most beautiful in the more-than-human world. Walton's book does not only *represent* nature. The process of reading and turning its pages *connects* the reader metonymically to natural processes: the rippling flow of the River Lea and the growth of the grass and water irises ('flags').

Clare, therefore, self-consciously rejected 'sublime' representations of the natural world in favour of what he described in his great poem 'The Flitting' (composed in 1832) as writing 'that mild and bland / breathes of green fields and open sky'.[93] Much of his nature writing is conducted at ground level, finding great significance in precise observations of

[91] This desire to conflate nature and nature writing is apparent in the same letter when he states that 'all nature is poetic' (p. 33).

[92] Clare's editors have identified this as a stone curlew, although in that case the reports he conveys that it could be heard 'at all seasons of the year' are incorrect, as they are summer migrants (p. 35).

[93] John Clare, *Poems of the Middle Period*, ed. Eric Robinson, David Powell and P. M. S. Dawson, 5 vols. (Oxford: Clarendon, 1996–2003), vol. 3, pp. 479–83, ll. 161–2.

apparently unimportant natural objects such as birds' nests and 'Pooty [snail] shells':

> I love to look on nature with a poetic feeling which magnifys the pleasure I love to see the nightingale in its hazel retreat & the cuckoo hiding in its solitudes of oaken foliage & not to examine their carcasses in glass cases yet naturalists & botanists seem to have no taste for this poetical feeling they merely make collections of dried specimens classing them after Leanius into tribes & familys & there they delight to show them as a sort of ambitious fame ... tho I feel as happy as they can in finding a new species of field flower or butter flye which I have not seen before yet I have no desire to dry the plant or torture the Butterflye by sticking it on a cork board with a pin. (pp. 38–9)

As has often been pointed out, many of Clare's poems begin with 'I love'; in this prose passage too, the affective element of ecological encounter is emphasised. 'Magnifys' is a kind of pun: pleasure, love and creativity come from the ability to change one's scale of perception so as to put oneself in the place of the observed creature. This is a different type of magnification than that required by the taxonomist: Clare's birds tend to be glimpsed sympathetically rather than scrutinised from a position of power. He is not interested in studying decontextualised specimens – a practice he presents as cruel and destructive – but in observing dynamic living creatures within their environment.[94] Like White and Smith, he is fascinated with birds as domestic creatures, both in their own right, and in their capacity to live with humans. Clare's seventh natural history letter is devoted to robins, wrens, (house) martins and hedge sparrows (dunnocks) and emphasises how they are treated as important members of the community; given such 'veneration' that there are 'many people who will not suffer their nests to be destroyed' (p. 52). He notes that 'the Robin seems to be fond of the company & haunts of man' and recalls one that used to visit his house over three winters and could be identified by 'a white scar on one of the wings' (p. 52). The image of the bird as a close acquaintance – even a family member – once again emphasises personal connection rather than the distanced analysis of the taxonomist.

Clare did not have access to the modern concept of 'ecosystem', but it is clear that he had an understanding of Helpston and its environs as constituted by a complex set of entanglements between people, birds,

[94] A good example of this dynamism is the sonnet 'Emmonsails Heath in Winter', as discussed in John Barrell's classic study of Clare, *The Idea of Landscape and the Sense of Place* (Cambridge: Cambridge University Press, 1972), pp. 155–62.

animals, plants, objects and landscapes. Even more so than Smith and White, Clare was concerned with the localised meanings and affinities associated with birds, animals and plants, rather than any universal taxonomy. As Theresa M. Kelley has argued in a brilliant discussion of Clare's botany, his rejection of the Linnaean system was strongly connected to his resistance to enclosure and his scepticism about grammatical rules.[95] Clare was fascinated by the 'instability and variability of common names' and their capacity to slip away from the taxonomist's 'fixed positions in a systematic grid'.[96] In Clare's nature writing, the reader encounters a multitude of vernacular names (often several for a single species) that go beyond mere representation by offering metonymic connection: for example, 'the long taild Titmouse called with us Bumbarrel & in yorkshire pudding bags & feather pokes' (p. 48). He relishes the richness of these localised names, and is particularly fascinated by instances of onomatopoeia. For example, in a note on the 'butter bump', Clare describes 'a thing that makes a very odd noise morning & evening ... it is difficult to describe what it is like its noise has procurd it the above name by the common people' (p. 89) this is the booming of the bittern, an elusive wading bird that is much easier to hear than to see. Appropriately, most of the note attempts to describe its distinctive call through a shifting set of analogies. 'The first part of its noise', Clare states, 'is an indistinct muttering sort of sound very like the word butter uttered in a hurried manner' (p. 89). Clare's language here does not only attempt to describe the call but also to replicate it through the internal rhyme of mutter-butter-utter. We are told that 'bump comes very quick after' and 'bumps a sound on the ear as if echo had mockd the bump of a gun' (p. 89). But Clare does not stop there; we are further informed that 'putting ones mouth to the bung hole of an empty large cask & uttering the word "butter bump" sharply would imitate the sound exactly' (p. 89). A person can sound like a bittern just as the best nature writing can replicate the more-than-human world. Like Smith, Clare attends to the cultural narratives that attach themselves to birds: as a child, he was told that the call came from 'a bird larger than an ox that coud kill all the cattle in the fen if it choose & destroy the villager likewise but that it was very harmless & all the harm it did was the drinking so much water as to nearly empty the dykes in summer & spoil the rest so that the stock coud scarcly drink what it left' (p. 89). At the end of the note there is only a brief description of the bittern 'on the wing', which reflects the fact that it was experienced much more through sound and story than

[95] Kelley, *Clandestine Marriage*, chapter 5. [96] *Ibid.*, p. 128.

through sight. Clare's approach is therefore the opposite of a taxonomical one, based on multidimensional localised experience, rather than the purely visual observation of a static specimen.

Clare ends with the observation that 'there is a great many of these birds on whittlesea mere and their noise is easily heard in a morning on the London road which is some miles distant' (p. 90). Around the middle of the nineteenth century, Whittlesea Mere was drained; one of the last fens to be destroyed for the purpose of agricultural 'improvement'. A generation later, the bittern was extinct in the British Isles. It returned in the twentieth century and since then its numbers have waxed and waned, but certainly have never got close to what would have been present in Clare's lifetime (the species is currently doing relatively well with 198 males counted in 2019).[97] Even if not for the traffic noise, one would not expect to hear bitterns while travelling along the 'London road' – that is, the A1. If Clare's writing often looks back to a time of lost plenitude associated with childhood, his present-day readers may experience a further temporal dislocation. Nature writing can function as elegy even when it does not construct itself as such. When Clare asks, with respect to the distinctive *crex crex* call of the 'land rail' (corncrake), 'Were [*sic*] is the school boy that has not heard that mysterious noise which comes with the spring in the grass' (p. 49), he could not have imagined a situation in which, due largely to changes in agricultural practice, the bird has been reduced to a tiny number of breeding pairs in the United Kingdom, and largely confined to the Scottish islands.

One of the most striking examples of Clare's nature writing is the long list of his observations of birds, which Grainger dates to 1825–6 (p. 118). As Grainger points out, the list is 'packed with detailed information' and 'memorable vignettes' that reveal Clare's remarkable capacity for observation (p. 120) and his knowledge of local ornithology. Even allowing for a degree of hyperbole, it presents an abundance of bird life that would be hard to imagine now in Britain outside landmark nature reserves. Lapwings (peewits) are certainly not 'as common as crows' in present-day Northamptonshire, and birds with which he is clearly familiar like wrynecks, quails, red-backed shrikes, whinchats and tree pipits would be difficult or impossible to find. Changing environmental conditions bring gains as well as losses. For example, Clare would not have encountered egrets, which are now increasingly common in England (probably due to

97 www.countryfile.com/wildlife/birds/britains-bittern-population-at-a-record-high/ (accessed 31 August 2020).

climate change). However, the dramatic overall decline of wildlife populations, globally and within in Britain, is well documented. This is known as 'defaunation': a less headline-grabbing term than extinction, but a more widespread and insidious process.[98] Clare did have a sense of the loss of flora and fauna created by enclosure, as well as broader trends, remarking at one point on the decline of nightingales, which 'like the Martins & Swallows & other birds of passage they seem to diminish but for what cause I know not' (p. 38). But he would not have been able to imagine the deracinated agricultural landscapes that now comprise much of Britain.

Clare's fascination with bitterns and corncrakes derived in part from the fact that they are much easier to hear than to see. He was similarly intrigued by another elusive bird with a distinctive call: the nightjar. 'The Fern Owl or Goat sucker or Night jar or night hawk', he notes, 'while several more or's might be added no doubt to fresh names is a curious bird' (pp. 32–3). These multiple names reveal the fluid identity of the bird, its importance in folklore and its resistance to taxonomic classification. Clare describes a specimen he has seen in his friend Artis' collection (p. 34), but he is clearly much more interested in what it means to encounter the living creature:

> They are found about us in summer on a wild heath calld Emmingsales & I believe that is the only spot which they visit they make an odd noise in the evening beginning at dew fall & continuing it at intervals all night it is a beautifull object in Poetic Nature—(nay all nature is poetic) from that peculiarity alone one cannot pass over a wild heath in a summer evening without being stopt to listen & admire its novel & pleasing noise. (p. 33)

As with the bittern, Clare sought to connect with the nightjar through mimicry; its call, he notes, 'is a trembling sort of crooing sound which may be nearly imitated by making a crooing ... noise & at the same time patting the finger before the mouth to break the sound like stopping a hole in a German flute to quaver a double sound on one note' (p. 33). By 'crooing', Clare probably means 'cooing'; in any case, doing what he suggests results in a sound that does not much resemble that of a nightjar. Neither, perhaps, does Bewick's suggestion that its call is similar to the sound made by a 'spinning-wheel', or Smith's description in the sonnet to which Clare refers that 'the shrieking night-jar sail on heavy wing'.[99] All

[98] See Hillary S. Young, Douglas J. McCauley, Mauro Galetti and Rodolfo Dirzo, 'Patterns, Causes, and Consequences of Anthropocene Defaunation', *Annual Review of Ecology, Evolution, and Systematics*, 47 (2016), 333–58.

[99] Curran (ed.), *The Works of Charlotte Smith*, vol. 14, p. 42.

three authors were well aware that inadequacy is built into nature writing because it can never be the equivalent of an actual encounter. But, at its best, such writing brings us closer to mysterious creatures like the nightjar at the same time as celebrating them for all the ways in which they are not like us.

A passing reference in the same letter reveals an important difference between Clare's nature writing and White's: 'My Love rambles then made me acquainted with many of the privacies of night ... which she seemd wishing to keep as secrets' (pp. 34–5). Even if White had made nocturnal 'Love rambles' (which seems unlikely), he would not have written about them in *Selborne*. Clare's nature writing is autobiographical, emotional and affective to a much greater extent than White's. It is also much more concerned with personal liberty. The ability to wander freely was vital to Clare's creativity, but was also a frequent cause of anxiety.[100] In a classic essay, John Goodridge and Kelsey Thornton analyse the significance of literal (as a pedestrian) and metaphorical (as a labouring-class poet) trespass in Clare's writings. They show that there was good reason to be frightened of trespassing during Clare's lifetime, given the draconian Game Laws and the often violent conflicts between labourers and gamekeepers. Such anxieties, they argue, contribute to the prevalence in Clare's writings of 'the imagery of land boundaries, and of the two functions of boundaries, to enclose, and to exclude'.[101] Helpston's enclosure also imprinted Clare's nature writing with a sense of loss and disenfranchisement, even if he also benefitted from the process through finding substantial employment within enclosure gangs.[102] For example, his remarkable poem, 'The Mores', describes what may well be Emmonsales Heath as perceived by Clare as a boy.[103] Once, there was no 'fence of ownership ... / To hide the prospect of the following eye' (ll. 8–9). Since enclosure, however, 'this sweet vision' (l. 14) has 'faded all' (l. 16), along with the 'commons wild & gay' (l. 39). When he states that 'these paths are stopt' (l. 65), the barriers are imaginative as well as topographical. What was once common land is now sealed off by tyrannical signs of ownership. Clare's nature writing is informed by personal, ecological and communal losses; paradoxically, the

100 Higgins, *Romantic Englishness*, chapter 4.

101 John Goodridge and Kelsey Thornton, 'John Clare: The Trespasser', in Hugh Haughton, Adam Phillips and Geoffrey Summerfield (eds.), *John Clare in Context* (Cambridge: Cambridge University Press, 1994), pp. 87–129.

102 For nuanced discussions of Clare and enclosure, see Castellano, *Ecology*, chapter 6, and Kövesi, *John Clare*, chapter 1.

103 Clare, *Poems of the Middle Period*, vol. 2, pp. 347–50. Line references are in the text.

internal colonisation of his parish landscape, despite being psychologically damaging, also generates a troubled creativity and a strong sense of connection with the nonhuman world.

Like Clare, James Hogg brought a distinctive labouring-class sensibility to his accounts of rural life and was both a distinguished poet and prose writer who had a complex relationship with metropolitan literary culture. His concern was with the culture and traditions of the Scottish borders; he was more interested in agriculture and cultural landscapes than the wild creatures beloved by Clare. However, he can be usefully understood as a nature writer in that he articulates a mode of existence in which human and nonhuman forces intertwine, and which is simultaneously modern in sensibility and resistant to the forces of modernity. His two-part tale-essay 'Storms' (1819) encapsulates this articulation.[104] It opened a series of articles in *Blackwood's Edinburgh Magazine* under the overall title of 'The Shepherd's Calendar', which were later published in book form. As David Stewart suggests, Hogg's Shepherd's Calendar stories 'move between forms and genres, and between tradition and modernity. Stories expand and develop seemingly without the volition of the narrator and inhabit as many genres and moods as possible. His stories experiment with the limits of their form; he finds in them something of the essay's freedom to follow a path and respond to the impulse of the moment.'[105] Like much of Hogg's work, 'Storms' is highly self-conscious and plays with its own textuality and generic roots. It is about a relatively isolated rural life, but written for a predominantly urban audience and reflects on that tension, which was both spatial (metropolis vs. periphery) and temporal (modernity vs. tradition). It describes the local effects on a sheep-farming community of an unusually violent storm that affected Scotland in 1794 and killed many sheep and other animals, as well as some people. The events are grim, even tragic, but in the telling are mingled with the self-parodic comedy which was the stock-in-trade of Hogg and the coterie of Edinburgh writers with which he associated. The autobiographical narrator is somewhere between a pastoralist and an author, part of a group of 'young shepherds' who 'had formed themselves into a sort of literary society that met periodically' (p. 5). He sets off to the meeting, a twenty-mile walk over

104 'Storms' was first published in the April 1819 *Blackwood's Edinburgh Magazine* and reprinted by Hogg several times. We use the version (based on the manuscript) in James Hogg, *The Shepherd's Calendar*, ed. Douglas S. Mack (Edinburgh: Edinburgh University Press, 2002), pp. 1–21. Further references are included in the text.

105 David Stewart, 'Romantic Short Fiction', in Ann-Marie Einhaus (ed.), *The Cambridge Companion to the English Short Story* (Cambridge: Cambridge University Press, 2016), pp. 73–86.

rugged countryside in winter, 'with a flaming bombastical essay in my pocket, and my tongue trained to many wise and profound remarks' (p. 5). This is more than incidental detail; later in the essay, a mysterious relationship develops between the young Hogg's melodramatic writing and the tragic drama of the storm itself. As the narrator walks to his literary meeting, he observes an ominous scene that is proleptic of the catastrophe to follow:

> The wind which was rough on the preceding day had subsided into a dead calm; there was a slight fall of snow, which descended in small, thin flakes that seemed to hover and reel in the air as if uncertain whether to go upward or downward; the hills were covered down to the middle in deep folds of rime, or frost-fog; in the cloughs [gorges] that was dark, dense, and seemed as if it were heaped and crushed together; but on the brows of the hills it had a pale and fleecy appearance and altogether I never beheld a day of such gloomy aspect. (p. 5)

The flakes of snow may be 'uncertain', but the word implies a decision-making capacity; they are ascribed a kind of agency which presages the enormous nonhuman force about to be unleashed. Similarly, the 'heaped and crushed together' fog in the 'cloughs' is later echoed by the groups of sheep buried deep in the snow. And the 'fleecy' fog is a reminder of the danger not only to Hogg's own sheep, but also to a rural community absolutely dependent on sheep farming.[106] As a result, the narrator decides to return home, noting ruefully that 'my elaborate essay, and all my wise suppositions had come to nothing' (p. 6).

The account of the storm that follows combines visceral descriptions of its power and effects with moments of comic bathos. The narrator's anxious nocturnal wait for its arrival is leavened when he wanders down to a local ruin, in a moment reminiscent of Clare's 'Love rambles', with 'a supposition that one of the [servant] girls would follow me or else that I would see a hare both very unlikely events to have taken place on such a night' (p. 6). The storm terrifies him: 'so completely was the air overloaded with falling and driving snow that but for the force of the wind I felt as if I had thrust my arm into a wreathe of snow. I deemed it a judgement sent from heaven upon us' (p. 7). But when he hears his bunkmate Borthwick crying out 'in a shrill desperate tone of voice' having got stuck in the snow while trying to cross a 'large dunghill', he cannot 'refrain from laughing immoderately notwithstanding the dismal prospect that lay before us'. The

[106] Ron Broglio, *Beasts of Burden: Biopolitics, Labor, and Animal Life in British Romanticism* (Albany: State University of New York Press, 2017), pp. 60–1.

two shepherds and their master leave before dawn to try to rescue their sheep. Conditions are appalling, with the darkness, 'eddying wind' and deep snow combining to slow and confuse them. Hogg finds humour even in this potentially deadly situation as he describes falling headfirst over a precipice and getting caught between it and the snow: 'for all our troubles I heard Borthwick above convulsed with laughter'. These moments of comedy and the abnormal power of the storm give the story a surreal air, making it akin to some of Hogg's more obviously supernatural tales. Having rescued some of his own sheep, the narrator, in a state of terror and fatigue, attempts to reach home before dark, and experiences an 'optical delusion ... so like enchantment', thinking that he sees 'trees over my head flourishing abroad over the whole sky ... to the utmost verge of the visible heavens'. This reinforces his pre-modern interpretation of the storm 'as a great judgement sent on us for our sins', somehow connected to the 'strange phantasy ... effected by evil spirits' (p. 10). Thus, the first part of 'The Storm' ends firmly situated a long way from civilisation, both geographically and temporally.

The second part complicates the apparent dichotomy between country/past and city/future. News gradually reaches the narrator of the devastation wreaked by the storm, including the deaths of 'thousands of sheep' and a number of shepherds: 'the country was a charnel-house' (p. 14). His view of the storm as divine punishment is widely shared, and the innocent meeting of the literary society becomes an object of suspicion: 'when it came out, that so many men had assembled, in a lone unfrequented place, and busily engaged in some mysterious work at the very instance that the blast came on, no doubts were entertained that all had not been right there, and that some horrible rite, or correspondence with the powers of darkness had been going on' (p. 15). This is expressed in the measured, standardised language of Edinburgh literary culture; the same point is made more forcefully in the reported dialect speech of Hogg's friend John Burnet: 'It seems that a great squad o' birkies [fools] that ye are conneckit wi', had met that night at the herds house o' Ever Phaup, an had raised the deil among them' (p. 16). The narrator's embarrassed fellowship with these 'birkies' ('I looked on the whole fraternity as my brethren') is a self-reflexive reference to Hogg's uneasy membership of the literary coterie surrounding *Blackwood's Magazine*. The reference is emphasised a little later in the essay, when the narrator states that 'if the effects produced by the Chaldee Manuscript had not been fresh in the minds of the present generation, they could have no right conception of the rancour that prevailed' (p. 17). 'The Chaldee Manuscript' (1817) was a controversial satirical article published in *Blackwood's*, to which Hogg contributed, and that caused outrage among the Edinburgh literary community. Like the use of pastoral elegy in 'Storms', the 'Manuscript' retooled ancient tropes (biblical prophecy) in a definitively modern context. Hogg's heterogeneous inventiveness is apparent towards the end of 'Storms', when the narrator's credulous mother quizzes the servant from the house where the literary society met. As in the rest of the essay, the semi-comic dialogue is in a rich vernacular that contrasts with the more standardised language of the narrator.

The narrative's self-reflexive intertextuality is typical of the *Blackwood's* coterie, and continues at the conclusion with 'an extract from a poet for whose works I always feel disposed to have a great partiality' (p. 19). The extract is from Hogg's most celebrated poem, *The Queen's Wake* (1814), and represents an incursion into the tale of the sublime language of poetic personification: 'Who scalp'd the brows of old Cairngorm, / And scoop'd these ever yawning caves? / 'Twas I, the Spirit of the Storm!' (p. 19). Ron Broglio is right that Hogg 'reminds the reader of forces that will not be co-opted' into mechanisms of biopolitical power 'be they sheep, shepherds, dogs, witches, brownies, or fairies' (or, I would add, storms). But he is only half right when he states that 'Hogg's sheep stories sound as if they are of another place and time because they are; which is to say, they are another world and a different temporality from that of Edinburgh and even by extension the larger national interests.'[107] Hogg's writings, like Hogg himself, straddle the two worlds and reveal parallels between them through textual heterogeneity and allusion. Like the other texts examined in this chapter, 'Storms' is concerned with the power of nature in motion, and with the entanglements of nonhuman and human forces. But as in those texts, it is ultimately most interested in the power of nature writing itself, with its refusal to follow conventional narratorial or generic patterns, its dialectic of empirical observation and self-referentiality, and its ironic attempts to cross the border between text and world.

Conclusion: The World without Us

This chapter has emphasised the modernity of Romantic nature writing and its continuities with later works in the genre. There were also significant differences. Not all the writers we have discussed were as religiously orthodox as the parson-naturalist Gilbert White. But they were all broadly

[107] *Ibid.*, p. 63.

working within a tradition of Christian natural theology that understood God's presence as immanent within nature. This tradition would prove surprisingly resilient in subsequent decades (see Chapter 2), but would also come under increasing pressure from developing understanding of the long and volatile histories of landscapes and species. It was in the Romantic period that the idea of 'deep time' began seriously to challenge orthodox biblical chronology, even though attempts were often made to reconcile the two. New geological ideas had a powerful impact on Romantic prose and poetry and encouraged the more radical questionings of sceptics such as Byron and Percy Shelley.[108] They also influenced more apparently conservative authors. For example, Charlotte Smith's most ambitious poem 'Beachy Head' (1807) offers a nature writing tour de force, as it moves through geology, landscape aesthetics, ornithology, botany, national and local histories, politics, folklore, georgic, pastoral and autobiography. It views a wide prospect of time and space, a mode traditionally associated in poetry with masculine power, while complicating its assertion of authority through fragmentation and multivocality.[109] It begins by vividly imagining the 'Omnipotent' separating the British Isles from the continent of Europe; the conventional reference to divine intervention should not disguise Smith's willingness to countenance new ideas about past changes to the planet's topography. Later, the speaker reflects on the possible meanings of fossils ('the strange and foreign forms / Of sea-shells'), asking whether 'Nature' mimics 'fantastic shapes / Of bivalves and inwreathed volutes', or whether these were once living creatures in a very different landscape:

> Or did this range of chalky mountains,* once
> Form a vast bason, where the Ocean waves
> Swell'd fathomless? What time these fossil shells,
> Buoy'd on their native element, were thrown
> Among the imbedding calx: when the huge hill
> Its giant bulk heaved.[110]

Smith's footnote to 'chalky mountains' refers 'to an idea started by Mr White'; there was nothing controversial about these speculations in

[108] Higgins, *British Romanticism, Climate Change, and the Anthropocene*; O'Connor, *The Earth on Show*.

[109] Labbe, *Charlotte Smith*, chapter 5. For the poem and geology, see Kevis Goodman, 'Conjectures on Beachy Head: Charlotte Smith's Geological Poetics and the Ground of the Present', *ELH*, 81 (2014), 983–1006; Tobias Menely, 'Late Holocene Poetics: Genre and Geohistory in Beachy Head', *European Romantic Review*, 28 (2017), 307–14.

[110] Curran (ed.), *Works of Charlotte Smith*, vol. 14, pp. 165–6.

themselves, even if they were to open up dangerous possibilities. As in 'The Swallow', Smith asks natural-philosophical questions only to draw back from them as 'but conjecture / Food for vain theories, or vain dispute'.[111] Both poems assert the (female) nature writer's intellectual authority while also acknowledging epistemological and representational limits; the views from Beachy Head are 'mocking alike [to] / The Poet and the Painter's utmost art'.[112]

As we have seen, Romantic nature writing was not only concerned with the importance of the apparently insignificant, but also with catastrophic phenomena on a vast scale. The most daring writers of the period speculated about the transience of the human species itself, or even – as in Byron's poem 'Darkness' (1816) – universal destruction without any religious recompense. We have argued in this chapter that Romantic nature writing is not so much a discrete genre as a hybrid mode of experiential and philosophical engagement across different kinds of text. As such, we will perhaps be forgiven for ending with a work of fiction: Mary Shelley's novel about a global pandemic, *The Last Man* (1826). In one remarkable passage, the novel's narrator, Lionel Verney, contemplates the landscape around Windsor, as he and the other main characters plan to flee the plague-wracked British Isles:

> For the last time we looked on the wide extent of country visible from the terrace, and saw the last rays of sun tinge the dark masses of wood variegated by autumnal tints; the uncultivated fields and smokeless cottages lay in shadow below; the Thames wound through the wide plain, and the venerable pile of Eton college, stood in dark relief, a prominent object; the cawing of the myriad rooks which inhabited the trees of the little park, as in column or thick wedge they speeded to their nests, disturbed the silence of evening. Nature was the same, as when she was the kind mother of the human race; now, childless and forlorn, her fertility was a mockery; her loveliness a mask for deformity. Why should the breeze gently stir the trees, man felt not its refreshment? Why did dark night adorn herself with stars – man saw them not? Why are there fruits, or flowers, or streams, man is not here to enjoy them?[113]

Here the end of the world is conflated with the end of humanity. The crux of reading this passage is the degree to which the novel encourages Verney to be viewed ironically. Is his anthropocentrism suspect? After all, his questions might be compared to the famous one at the end of Percy

[111] *Ibid.*, p. 166. [112] *Ibid.*, p. 165.
[113] Mary Shelley, *The Last Man*, ed. Morton D. Paley (Oxford: Oxford University Press, 1994), p. 329.

Shelley's poem 'Mont Blanc' (1816) – 'And what were thou, and earth, and stars, and sea, / If to the human mind's imaginings / Silence and solitude were vacancy?' – and might similarly seem to imply that the world has no meaning without human perception.[114] To suggest that Shelley is ironising Verney's position would neatly fit her in with an anti-anthropocentric ethic with which many of us would be inclined to agree today. However, as Timothy Clark has pointed out, too many ecocritical readings expend too much labour showing that their texts reflect the ethical-environmental truisms that critics and their readers probably already hold.[115] There is no evidence that Verney's elegiac portrayal of a depopulated landscape is not to be taken seriously. The plague has achieved a kind of extreme picturesque by occluding all rural labour and transforming elite social institutions such as Eton into empty ruins. For Verney and Shelley, the flourishing of other forms of life, evident in the cawing of the rooks, does not make up for this loss.

Romanticism is sometimes blamed for the anthropocentric thinking most obviously apparent in Kant's assertion of the autonomy of the human subject in relation to nature.[116] This claim is not without basis, but Romantic nature writing could also involve decentring the human subject. This was apparent not only in its interest in the remarkable lives of apparently insignificant creatures, but also in its emerging sense of the vulnerability of the human species in relation to uncontrollable natural forces. Through its concern with nonhuman agency, Romantic nature writing at times challenged the apparent separation between nature and culture that characterises Latour's 'modern constitution'. This challenge was also enabled by writers' self-consciousness about the limitations of ecomimesis: the ways in which they understood that nature could not be mastered by human forms of thought. Later in the nineteenth century, Charles Darwin's shocking theory of evolution, with its emphasis on randomness, competition and extinction, would blow natural theology out of the water (see Chapter 2). In doing so, he was pushing against some aspects of Romantic nature writing, but simultaneously drawing on others. Nature writers still contend with this ambivalent inheritance today.

[114] Percy Bysshe Shelley, *Shelley's Poetry and Prose*, ed. Donald H. Reiman and Neil Fraistat, 2nd edition (New York: Norton, 2002), p. 101.

[115] Timothy Clark, *Ecocriticism on the Edge: The Anthropocene As a Threshold Concept* (London: Bloomsbury, 2015), p. 78.

[116] Nigel Clark provides an excellent discussion of Kantian thought in relation to intimations of human vulnerability in the eighteenth century: *Inhuman Nature: Sociable Life on a Dynamic Planet* (London: Sage, 2011), chapter 4.

CHAPTER 2

Victorian

Introduction

In the Victorian period, nature studies emerged as a popular pastime, accompanied by a publishing boom in accessible natural history texts for general readers. Such writing could overlap with art criticism, travel narrative, personal memoir and religious sermonising. It vacillated between the escapist idealism of the pastoral and the hands-on realism of the georgic. Its narrators were at once labourers, collecting specimens and observations with tireless industry, and leisured consumers, deriving aesthetic pleasure from wondrous landscapes. Systematic descriptions of environments, and authors' studies of them, were juxtaposed with rhetoric of the natural world as a 'fairyland' of amusing mystery and strangeness.[1] Such rhetoric served not only to attract readers unversed in natural sciences, but also to resist disturbing aspects of scientific rationalism and materialism. Prose focused upon natural environments and objects often sought to reconcile modern knowledge of physical nature with religious orthodoxy or at least some form of spirituality. Although no nature author of equivalent stature to Henry David Thoreau emerged in Victorian Britain, many writers imagined nature as having metaphysical significance

[1] See, for example, Arabella Buckley, *The Fairy-Land of Science* (London: E. Stanford, 1878). On natural history and imaginative escapism, see Lynn L. Merrill, *The Romance of Victorian Natural History* (Oxford: Oxford University Press, 1989) and Melanie Keene, *Science in Wonderland: The Scientific Fairy Tales of Victorian Britain* (Oxford: Oxford University Press, 2015). On the growth of popular natural history publishing in the period, see David Elliston Allen, *The Naturalist in Britain: A Social History*, 2nd edition (Princeton, NJ: Princeton University Press, 1994); Bernard Lightman, *Victorian Popularizers of Science: Designing Nature for New Audiences* (Chicago, IL: University of Chicago Press, 2009); Barbara T. Gates, 'Introduction: Why Victorian Natural History?', *Victorian Studies*, 35 (2009), 539–49.

and value in ways comparable to Thoreau and his Transcendentalist mentor Ralph Waldo Emerson.[2]

Central to this tendency was an idealist stance that regarded the natural world as ultimately inseparable from the human mind, summed up in William Wordsworth's dictum that our faculties 'half create' the world they 'perceive'.[3] For Victorians, haunted by the supposed excessive rationalism of the French Revolution, such idealism was often associated with deprecation of scientific efforts to explain nature rationally. Instead, authors such as John Ruskin presented purely logical, materialist inquiries as an impoverishment of nature that missed its spiritual aspects, manifested in aesthetic and emotional effects. Contrary to scientific representations, which claimed to capture nature via fixed mechanical laws, Ruskin emphasised nature's elusiveness, mutability and ultimate unknowability. This attitude was contiguous with a rhetoric that framed natural knowledge as intuitive revelations as well as logical inferences from hard facts. It also connected to a formal poetics that foregrounded the subjective impressions of wandering individuals immersed in natural environments.

Such *conservative idealism*, as we call it, cohered with the construction of nature studies as a vehicle of cultural refinement. Although this attitude could reaffirm the pastoral's division between leisured appreciators of natural landscapes and impoverished labourers within them, some authors from humbler backgrounds also blurred the lines between work and leisure, presenting nature appreciation as a democratic mode of self-improvement. Conservative idealism also sometimes encouraged a nascent ecological and humanitarian sensibility by ascribing value to wildlife beyond material expedients. However, placing the human mind at the centre of the universe necessarily limited the possibilities of ecological and animal-liberationist thinking. Some Victorian writing about nature also indexed a turn against idealism that grew through the century, framing wild environments and organisms as radically alien to the human mind. New narratives of the Earth's history, theories of evolution and materialist psychology all rendered nature not only strange, but estranged and estranging. Strangeness further featured in Victorian writing about nature via

[2] Although Emerson would be widely read, Thoreau's work remained little known in Britain until near the end of the century, when socialists such as Edward Carpenter began to promote it. See Peter Adkins, 'Transatlantic Dialogues in Sustainability: Edward Carpenter, Henry David Thoreau and the Literature of Simplification', in Wendy Parkins (ed.), *Victorian Sustainability in Literature and Culture* (New York: Routledge, 2018), pp. 51–68.

[3] William Wordsworth, 'Lines Composed a Few Miles above Tintern Abbey, On Revisiting the Banks of the Wye during a Tour. July 13, 1798', in *The Poems of William Wordsworth*, ed. Jared Curtis, 3 vols. (Cumbria: Humanities-Ebooks, 2011), I.372–6, p. 375.

clashing discourses of geographic identity. Victorian authors often divided natural objects into the native and foreign, but they also problematised this binary. Efforts to fold certain wildlife and landscapes into a narrative of national identity rooted in physical nature clashed with more flexible outlooks on place and belonging. This conflict paralleled uncertainties over where exactly to locate 'nature'. While some authors opposed their rural and coastal foci to Britain's urban centres, others blurred the lines between natural and rural landscapes.

This chapter discusses a wide range of authors, only two of whom fully fit the Thoreauvian mould of the 'nature writer' as a solitary wanderer who respects scientific facts while also regarding nature as sacred and revelatory. However, Richard Jefferies and William Henry Hudson followed a long line of writers who also blurred the boundaries between science, aesthetic appreciation and imaginative and religious reverie. As an immensely influential cultural critic and amateur naturalist, Ruskin warrants extended consideration. The intermingling of aesthetic and religious discourse with the scientific was pervasive in Victorian popular natural history, and works by Philip Henry Gosse, Charles Kingsley, Margaret Gatty, William Warde Fowler and Eliza Brightwen all demonstrate this hybridity. Leslie Stephen's writing on mountaineering in the Alps is similarly hybrid, reflecting self-consciously on the slippage between the facts of nature and its construction in the literary imagination.

Conservative Idealism and Nature As Revelation

Katey Castellano has argued that the environmental visions of Wordsworth and his intellectual allies drew on Edmund Burke's conservative 'intergenerational imagination', which emphasised respect for tradition against Enlightenment efforts to rationalise the world.[4] Such conservative idealism rejected scientific mechanicism, which tended to reduce the natural world to predictable processes that could be understood in purely physical terms without recourse to metaphysical causes or values. Constructing nature as a collection of objective, amoral facts, the mechanistic orientation invalidated inner experience as a way of knowing nature, rendering it a mere product of natural causes.[5] This conclusion had been exemplified in Julien Offray de La Mettrie's formulation of humans as

[4] Katey Castellano, *The Ecology of British Romantic Conservatism, 1790–1837* (New York: Palgrave, 2013), p. 19. See Chapter 1 for a discussion of Castellano's readings of Bewick and Clare.

[5] See Geert-Jan de Klerk, 'Mechanism and Vitality. A History of the Controversy', *Acta Biotheoretica*, 28 (1979), 1–10.

machines and Baron d'Holbach's view of the universe as simply matter in motion.[6] Conservative idealism can be understood as a reaction against such materialism, which became associated with radical politics through the French Revolution. It insisted that the physical world was ultimately subordinate to humans' inner intuitions, which proceeded from a higher spiritual plain beyond rational analysis.[7] This philosophy was articulated to Victorian audiences most influentially through Samuel Taylor Coleridge's concept of 'Imagination' and Thomas Carlyle's rhetoric of 'natural supernaturalism'. Their arguments would be echoed in much popular natural history writing that sought to reconcile materialist science with religious belief of one kind or another. This reconciliation was frequently figured through a poetics of revelation, which constructed nature studies as a partly spiritual exercise characterised by mysterious epiphanies and intimations of the divine or transcendental.

Coleridge and Carlyle's conservative idealism drew on German *Naturphilosophie*'s figurations of nature as a dynamic, vital totality in which matter and spirit interacted, as opposed to materialist mechanicism (see the discussion of 'nature in motion' in Chapter 1).[8] Coleridge described nature as a symbolic mirror of humanity's destined spiritual development, declaring, 'what the plant *is*, by an act not its own, and unconsciously – *that* must thou *make* thyself to *become*!'[9] Humans' ability to recognise such higher meanings in nature proceeded, he argued, not from 'mechanical' gathering of facts (a faculty he designated 'Fancy'), but intuitive 'Imagination'. Coleridge suggested that the mysteriousness of the latter faculty testified to its 'vital' character, linking humans to God, the source of all creativity.[10] Similarly, Carlyle's satire *Sartor Resartus* (1833–4) appealed for a vitalist science that would recognise nature's 'infinite depth'. Nature, Carlyle's fictional alter ego explained, was 'a Volume written in celestial hieroglyphs, in the true Sacred-writing; of which even Prophets are happy that they can read here a line and there a line'. Empirical science did not so much read this volume as 'pick out, by dexterous combination,

6 Julien Offray de La Mettrie, *Machine Man and Other Writings*, trans. and ed. Ann Thomson (Cambridge: Cambridge University Press, 1996); Baron d'Holbach, *The System of Nature*, trans. and ed. H. D. Robinson, 2 vols. (New York: Matsell, 1836).

7 On Victorian idealism, see Frank M. Turner, *Contesting Cultural Authority: Essays in Victorian Intellectual Life* (Cambridge: Cambridge University Press, 1993), pp. 322–61.

8 See Robert Richards, *The Romantic Conception of Life: Science and Philosophy in the Age of Goethe* (Chicago, IL: University of Chicago Press, 2002), pp. 1–16.

9 Samuel Taylor Coleridge, *Lay Sermons*, 3rd edition (London: Edward Moxon, 1852), pp. 76, 33.

10 Samuel Taylor Coleridge, *Biographia Literaria* (Cambridge: Cambridge University Press, 1920), p. 178.

some Letters in the vulgar Character, and therefrom put together this and the other economic Recipe'. To apprehend nature's higher meanings, to understand it as more than a 'Domestic-Cookery Book', required, in Carlyle's view, an intuitive response to nature, involving moral and aesthetic feelings and imagination, as well as logical inferences.[11] Scientific practitioners did not necessarily disagree. The astronomer and mathematician William Whewell, for example, argued forcefully against the Utilitarian John Stuart Mill that empirical research was guided by innate ideas.[12]

Such idealism fitted with a general tendency in early to mid-Victorian popular natural history to construct nature studies as a combination of methodical reasoning and religious intuitions. Rather than staking their beliefs upon natural evidence, authors generally presented their studies as appreciations of a divine design that did not need to be proved empirically.[13] Amy King describes an attitude of 'reverent empiricism' in such writing, which framed close attention to minute physical details as a pathway to general, spiritual truths.[14] This attitude entrenched a division between the methodical work of surveying environments and examining objects, and the imaginative intuition of a higher, spiritual reality. The reverent student of nature could be at once a hard-headed stickler for facts and a dreamer chasing intimations of immortality. In this way, his dual character reflected Carlyle's distinction between 'Manufacture' and 'Creation':

> [W]hat is mechanical lies open to us: not what is dynamical and has vitality. Of our Thinking, we might say, it is but the mere upper surface that we shape into articulate Thoughts; underneath the region of argument and conscious discourse, lies the region of meditation; here, in its quiet, mysterious depths, dwells what vital force is in us . . . Manufacture is intelligible, but trivial: Creation is great, and cannot be understood.[15]

11 Thomas Carlyle, *Sartor Resartus* (Oxford: Clarendon, 1913), pp. 184–5.

12 William Whewell, *Of Induction, With Especial Reference to Mr. J. Stuart Mill's System of Logic* (London: J. W. Parker, 1849), pp. 1–24.

13 Aileen Fyfe, *Science and Salvation: Evangelical Popular Science Publishing in Victorian Britain* (Chicago, IL: University of Chicago Press, 2004), pp. 3–7; Lightman, *Victorian Popularizers*, pp. 24, 81–2; Jonathan Topham, 'Science, Natural Theology, and the Practice of Christian Piety in Early Nineteenth-Century Religious Magazines', in Geoffrey Cantor and Sally Shuttleworth (eds.), *Science Serialized: Representations of the Sciences in Nineteenth-Century Periodicals* (Cambridge, MA: MIT Press, 2004), pp. 37–66.

14 Amy M. King, *The Divine in the Commonplace: Reverent Natural History and the Novel in Britain* (Cambridge: Cambridge University Press, 2019).

15 Thomas Carlyle, *Characteristics* (Boston: James R. Osgood and Company, 1877), pp. 10–11.

Like Coleridge, Carlyle viewed artistic creation as an echo of divine creativity, a higher power that worked through humans, beyond their rational understanding. Such thinking encouraged some Victorians to challenge the opposition of science and poetry, arguing that science itself could be a form of 'poetry', or at least complementary to it. Thus, in *The Poetry of Science* (1848), the mineralogist Robert Hunt wrote that natural facts had the 'power of exalting the mind to the contemplation of the Universe'.[16] Far from dispelling belief in the miraculous, Hunt insisted, nature studies (when done rightly) enhanced humans' sense of the mysterious spiritual power that lay behind phenomena.

Conservative idealism often manifested in prose about the natural world via a rhetorical construction of nature as a scene of revelations beyond scientific methods. Hunt commented that 'The revelation of truth is the Poetry of Science', defining 'The True' as part of a Platonic unity of 'the Beautiful' and the Good.[17] Similarly, in his memoir of geological researches, *The Old Red Sandstone* (1841), Hugh Miller presented rock formations and fossils as divine artworks forged to stimulate thoughts of the creator. Examining the outlines of extinct fish, Miller declares, 'There is unity of character in every scale, plate, and fin – unity such as all men of taste have learned to admire', echoing 'the masculine Doric, the chaste and graceful Ionic, [and] the exquisitely elegant Corinthian' styles of classical architecture. Miller presents appreciation of this natural beauty as a spiritual exercise:

> The wonders of geology exercise every faculty of the mind – reason; memory, imagination . . . We speak of the infinity of Deity – of his inexhaustible variety of mind . . . It is well to be brought to feel, if not to conceive of it . . . Succeeding creations, each with its myriads of existence, do not exhaust Him. He never repeats himself. The curtain drops, at his command, over one scene of existence full of wisdom and beauty; it rises again, and all is glorious, wise, and beautiful as before.[18]

Miller's adulation at the beauty of the fossil record downplays the disruptive implications of geological deep time and species extinctions. Geology was forcing theologians to place a newly figurative gloss on the Genesis narrative and to reconceptualise creation as a succession of epochs through the Earth's long history. However, Miller's emotionally charged description of fossils avoids their problematic aspects, interpreting them as aids for

[16] Robert Hunt, *The Poetry of Science* (London: Reeve, Benham, and Reeve, 1849), p. vii.
[17] *Ibid.*, pp. xvii–xx.
[18] Hugh Miller, *The Old Red Sandstone* (Boston: Gould and Lincoln, 1851), pp. 88, 93–4.

envisaging God. Like other artworks, Miller's natural world invites the observer to sympathise with its maker and infer his intentions. Miller's theatrical imagery further suggests that nature's secrets are miraculously shown rather than logically inferred.

Ruskin's art criticism similarly presented the natural world as a collection of moral-aesthetic revelations designed to elicit specific reactions from the human soul. 'Landscape', he wrote in the third volume of *Modern Painters* (1843–60), contained 'the grounds of noble emotion', which exceptionally sensitive seers such as J. M. W. Turner captured in their paintings. Ruskin privileged intuition over reason in his depiction of nature's beauty as a form of truth conveyed through imaginative vision rather than self-conscious logic. 'The greatest thing a human soul ever does in this world is to *see* something', he wrote. 'Hundreds of people can talk for one who can think, but thousands can think for one who can see. To see clearly is poetry, prophecy, and religion, – all in one.'[19] Strongly influenced by evangelical theories of typology, which approached history as a prophetic text, Ruskin imagined the natural world cryptically foreshadowing humanity's fall and salvation.[20] Different animals, he argued, could symbolise different vices and virtues. Hence, 'the earnest ant and unwearied bee' arouse affection, while 'we look with hate on the foulness of the sloth, and the subtlety of the adder, and the rage of the hyena'.[21] He even found spiritual allegories in the cyclical transformations of rocks. These changes, Ruskin claimed, foreshadowed 'the various states of mankind', from 'that which is powerless through total disorganization' to 'the perfect state of brotherhood and strength in which each character is clearly distinguished, separately perfected, and employed in its proper place and office'.[22] The virtues on which humans built civilisation seemed to be

[19] John Ruskin, *Works*, ed. Edward T. Cook and Alexander Wedderburn, 39 vols. (London: George Allen, 1903–12), V., pp. 387, 333.

[20] On Ruskin and Victorian typology, see George P. Landow, *The Aesthetic and Critical Theories of John Ruskin* (Princeton, NJ: Princeton University Press, 1971), pp. 329–56. On Ruskin's scriptural interpretation of nature, see also Stephen C. Finley, *Nature's Covenant: Figures of Landscape in Ruskin* (University Park: Pennsylvania State University Press, 1992); Lars Spuybroek, *The Sympathy of Things: Ruskin and the Ecology of Design*, 2nd edition (London: Bloomsbury, 2016); and Mark Frost, 'John Ruskin, Environment and the Ecological Impulse', in Laurence W. Mazzeno and Ronald D. Morrison (eds.), *Victorian Writers and the Environment: Ecocritical Perspectives* (New York: Routledge, 2016), pp. 13–28. On his complex attitude(s) towards science, see Frederick Kirchoff, 'A Science against Sciences: Ruskin's Floral Mythology', in U. C. Knoepflmacher and G. B. Tennyson (eds.), *Nature and the Victorian Imagination* (Berkeley: University of California Press, 1977), pp. 246–58; Francis O'Gorman, 'Ruskin's Science of the 1870s: Science, Education, and the Nation', in Dinah Birch (ed.), *Ruskin and the Dawn of the Modern* (Oxford: Oxford University Press 1999), pp. 33–55.

[21] Ruskin, *Works*, IV, p. 154. [22] *Ibid.*, VI, p. 133.

prefigured and preached by the natural environment. However, such meanings were elusive and intuitive, in contrast to the definite facts of materialist science.

The approach to nature as a scene of divine moral teachings drew on the tradition of moral sense theory, which posited that morality derived from innate feelings.[23] The parson-naturalist Charles Kingsley remarked in his book of seaside studies *Glaucus: Wonders of the Shore* (1855): 'How easily a man might, if he would, wash his soul clean . . . by going out to be alone with God in heaven, and with that earth which He has given to the children of men.' Natural or rural environments could be imagined as nurturing the moral sense because they were designed by God, in contrast to artificial towns and cities. This pastoral logic caused Kingsley to contrast the 'paradise' of the Devon coast, 'furred to the water's edge with rich oak woods', with the urban 'infernos of this sinful earth', 'the festering alleys of Bermondsey and Bethnal Green', where 'cholera' went hand in hand with 'carelessness, and laziness, and greed'.[24] A campaigner in the sanitation movement, Kingsley depicted urban epidemics as caused by city dwellers' alienation from nature, which degraded their desire for cleanliness, along with other virtues. Escaping into natural landscapes counteracted this degradation, he claimed, developing correct morals and tastes together via God's 'living and life-giving garment of perpetual health and youth', manifested in 'grass, herb, and tree'.[25] Kingsley's elision of moral conscience with 'health' figures the revelations of the natural world as intuitive rather than rational, similarly as Ruskin elided morals with appreciation of nature's beauty.

Kingsley further emphasised the intuitive aspects of nature studies by presenting the activity as a way of resting the faculty of reason, which was overworked by modern life and secular theorising. Excessive rationalism was a pathological temptation, he suggested, from which reverent natural history provided a 'healthy' distraction. Rambling is often considered relaxing, he writes. Yet when intelligent young men go for a walk 'without an object', they 'fall on arguing and brainsbeating on politics or metaphysics . . . and return with their wits even more heated and tired

[23] On moral sense theory in Victorian culture, see Linda Dowling, *The Vulgarisation of Art: The Victorians and Aesthetic Democracy* (Charlottesville: University of Virginia Press, 1996).

[24] Charles Kingsley, *Glaucus: Wonders of the Shore* (London: Macmillan and Company, 1873), pp. 221–2, 173–4. For ecocritical readings of Kingsley, see Christopher Hamlin, 'Charles Kingsley: From Being Green to Green Being', *Victorian Studies*, 54 (2012), 255–81; Akemi Yoshida, 'Ecotheological Morality in Charles Kingsley's *The Water-Babies*', in Dewey W. Hall (ed.), *Victorian Ecocriticism: The Politics of Place and Early Environmental Justice* (London: Lexington Books, 2017), pp. 111–26.

[25] Kingsley, *Glaucus*, p. 176.

than they were when they set out'. By supplying 'an object' for such walks, Kingsley explains, geology, botany and zoology can teach such youths to avoid morbid abstraction, riveting their minds to concrete realities. Nature studies medicate the mind, he continues, by teaching it 'to find wonder in every insect [and] sublimity in every hedgerow'.[26] Kingsley's rhetoric stresses the intellectual humility of the naturalist, who views nature as a mystery beyond his power to grasp fully. Confronted with the genius of the creator, he can only feel admiration and awe. Unlike the restless, ambitious man of science who seeks to conquer nature, emblematised in Mary Shelley's *Frankenstein* (1818), Kingsley's naturalist is content to *not* know all, trusting in God's superior wisdom (compare the discussion of Charlotte Smith's poetic personae in Chapter 1).

The uncritical, medicinal view of nature studies is foregrounded even more in *A Naturalist's Rambles on the Devonshire Coast* (1853) by Philip Henry Gosse, dubbed 'the David Attenborough of his day' due to his prolific output and immense popularity.[27] The narrative opens with Gosse's wife prescribing a rural retreat to the author, who has been made 'seriously ill' by overwork and their 'dreary' artificial city environs. Having collected many specimens of sea creatures along the coast, he houses them in an aquarium. After some failed experiments, Gosse succeeds in keeping the creatures alive, and notes repeatedly their 'health and vigour'. Their condition mirrors his own restoration thanks to the bracing littoral landscapes and the pleasure of nature studies, which he portrays as more play than work. Gosse calls his months on the coast 'one continuous, unflagging holiday' and invites his readers likewise 'to drink of this ever flowing stream of pure water', that is to say, the 'delightful' experience of nature studies.[28] His imagery of rest, amusement and hydration figures natural history as an activity that simultaneously stimulates and relaxes the mind. While the student of nature discovers endless varieties of living forms, the religious-theoretical framework through which he processes them remains stable, almost to the point of automaticity. Gosse presents his studies as less a demonstration of divine creation than simple appreciation of it. Of a bustling group of zoophytes and crustaceans in a rock pool, he

[26] *Ibid.*, pp. 53–5.

[27] Stephen Jay Gould, *The Flamingo's Smile* (London: Penguin, 1985), p. 100. For ecocritical analyses of Gosse, see Adam Nicol, 'The Ecology of Science and Slavery in Philip Henry Gosse's *Letters from Alabama*', *Prose Studies*, 35 (2013), 189–201; Sue Edney, 'Philip Henry Gosse, Newfoundland, and the Unveiling of Wonders', in Hall, *Victorian Ecocriticism*, pp. 127–48.

[28] Philip Henry Gosse, *A Naturalist's Rambles on the Devonshire Coast* (London: John Van Voorst, 1853), p. 436.

writes: 'We need not then ask for what purpose God has made so great a variety of creatures of no apparent benefit to man.'[29] Gazing into the rock pool, Gosse explains, illustrates God's 'beneficent care', not only for the animals that live by nature's balanced systems, but also for humans, who find moral lessons in them. Studying nature is for Gosse an exercise in reassurance, quieting the questioning mind by keeping it busy. Marine studies emerges as an antidote to temptations both bodily and mental, as Gosse makes clear in the preface when contrasting the naturalist with the seaside's 'idle pleasure-seekers'.[30] Such representations of nature studies as medicinal trace a bridge between Jean-Jacques Rousseau's restorative botanising (see Chapter 1) and the 'recovery narrative' common to much recent nature writing (see Chapter 4).

As the cultural authority of secular science grew through the period via the efforts of campaigners such as Thomas Henry Huxley, authors of the conservative-idealist persuasion increasingly differentiated their writing from specialist work. In 1860, Gosse contended that 'There are more ways than one of studying natural history', contrasting 'Dr Dryasdust's' fussy measurements of museum specimens and fact-gathering in the field with his own 'poet's way' that 'deals, not with statistics, but with the emotions of the human mind . . . which are made energetic by the creatures around him'.[31] In contrast to earlier assertions that all science was necessarily poetic or romantic, Gosse now limited the scope of his idealism to a special genre of natural history writing. Similarly, in *Love's Meinie* (1873–81), a lecture series on birds, Ruskin portrayed his aesthetic view of nature as a countercheck to secularists' focus on material facts and economic models of survival. The latter approach, he suggested, was an impoverishment of nature, not so much wrong as myopic. He attacked Darwinian sexual selection as 'the work of human bats; men of semi-faculty or semi-education', whose 'one-sided intensity' rendered them 'more or less incapable of so much as seeing, much less thinking about colour'. Studies of nature without a sense for good or beauty, he declared, were, above all 'vulgar' – the ultimate condemnation in a society that revered self-culture and refinement.[32] Ruskin's choice of words also points to his identification of such science with the industrial-capitalist order he raged against and the political-economic theory that justified it. In *Fors Clavigera* (1871–84), he blamed political economy for draining human

[29] *Ibid.*, p. 207. [30] *Ibid.*, p. v.
[31] Philip Henry Gosse, *The Romance of Natural History*, 2 vols. (London: James Nisbet and Company, 1871), I, p. v.
[32] Ruskin, *Works*, XXV, p. 264.

society of moral feeling and value, reducing life to mathematical calculations. Ruskin framed Darwin and his allies as accomplices to this degradation, complaining that 'modern science' seeks 'to show that there is no such thing as a flower . . . No such thing as a man, but only a Mechanism'. This reductive orientation, Ruskin wrote, 'disregarded the life and passion of the creature, which were its essence'.[33] Like the capitalist who knew the price of everything and the value of nothing, Ruskin's philistine scientist recognised only the vulgar, physical qualities of things.

Ruskin particularly critiqued science's vulgarity by attacking its habits of nomenclature, which, he claimed, stripped nature of its emotional-spiritual resonances accrued through centuries of human experience and stored in the national language. Ruskin subscribed to a vitalist model of philology, embodied in the theories of his friend Friedrich Max Müller. This paradigm constructed languages as organisms of a sort that matured through history, and verbal meaning as a kind of spiritual power concentrated in the etymologies of words.[34] Modern scientific nomenclature, by contrast, took its cue from John Locke's formulation of words as arbitrary 'counters' for external 'things'.[35] On this basis, researchers had long produced rival systems of nomenclature, each arguing that theirs distinguished more precisely between life's forms. Their assumption that words could be made to correspond so simply to things encapsulated for Ruskin the 'arrogance' of 'materialism'. Ruskin linked the scientific obsession with stamping new names upon bird varieties with base acquisitiveness. Modern ornithological papers, he complained, sacrificed observation of actual birds for bickering over what to call them.[36]

Against this possessive philistinism, Ruskin advocated 'the simplest and most descriptive nomenclature', both to make ornithology more inclusive and to bind it to a heritage of emotional reactions to birds. He advised his audience to 'cherish the old Saxon-English and Norman-French names of birds . . . never despising even the rudest or most provincial forms; all of them will, some day or other, give you clue to historical points of interest'.[37] Ruskin demonstrates this point by tracing the name

[33] *Ibid.*, XXVII, pp. 83–4.

[34] On Müller's influence upon Ruskin, see Edward T. Cook, *The Life of John Ruskin*, 2 vols. (New York: Haskell House, 1968), II, p. 157. On Victorian 'language vitalism', see Will Abberley, *English Fiction and the Evolution of Language, 1850–1914* (Cambridge: Cambridge University Press, 2015), pp. 14–18.

[35] John Locke, *An Essay Concerning Human Understanding* (London: William Tegg and Company, 1877), p. 525.

[36] Ruskin, *Works*, XXV, p. 20.

[37] *Ibid.*, p. 22. Compare the discussion in Chapter 1 of John Clare's interest in vernacular names.

buzzard back to medieval and early modern cultures of falconry and literary associations of the bird with vain amusements. Keeping the old, common word, he concludes, fills one's studies of the bird with 'freshness and value' in contrast to the empty, ahistorical names that 'anatomists blunder out . . . from their Latin dictionaries'.[38] Ruskin's opposition frames scientific nomenclature as artificial, in contrast to the poetic heritage of language, which appears to have unfolded naturally without human planning. Ruskin thus frames his poetic language as uniting humans with nature rather than separating them from it. Raised as an evangelical, Ruskin suffered a protracted crisis of faith from 1858 onwards as scientific developments undermined biblical literalism. Works such as *Love's Meinie* can be read as efforts to work through this crisis, seeking to preserve the mysterious and transcendental in nature against efforts to reduce it to mechanical laws. Humans' immanence in nature, he suggested, barred them from full mastery or understanding of it.

While Ruskin would ultimately sanitise his pantheism as an idiosyncratic Christianity, Jefferies developed similar ideas into an explicitly pagan nature worship.[39] This attitude is expounded most fully in his memoir *The Story of My Heart* (1883), a strange, unclassifiable book that Jefferies described as 'a confession', which unfolds as a kind of spiritual autobiography. Jefferies recalls having long experienced mystical epiphanies whenever he ventured alone into natural landscapes. Away from society, on a grassy hill, he writes, 'at every step my heart seemed to obtain a wider horizon of feeling; with every inhalation of rich pure air, a deeper desire'. He continues, 'Lying down on the grass, I spoke in my soul to the earth, the sun, the air . . . By all these I prayed; I felt an emotion of the soul beyond all definition; prayer is a puny thing to it.'[40] Nature's revelations for Jefferies are of a higher plain of existence which is never clearly defined. Like earlier, Christian naturalists, who stressed the incomprehensible greatness of the creator, he envisages the universe as sublimely overwhelming. In contrast to such predecessors, however, he imagines that humans might, through continuous effort, expand their view of the world, becoming godlike instead of simply submitting to a god. 'Let my soul become

[38] *Ibid.*, p. 23.

[39] On Richard Jefferies and nature, see Brian Morris, *Richard Jefferies and the Ecological Vision* (Victoria: Trafford, 2006); Rebecca Welshman, 'Literature and the Ecological Imagination: Richard Jefferies and D. H. Lawrence', *Victorian Network*, 3 (2011), 51–63; Ronald D. Morrison, 'Agriculture and Ecology in Richard Jefferies' Hodge and His Masters', in Mazzeno and Morrison, *Victorian Writers and the Environment*, pp. 205–19.

[40] Richard Jefferies, *The Story of My Heart* (London: Longmans, Green and Company, 1883), pp. 3–5.

enlarged', he recalls asking the earth, sun and sky. 'I desire a greatness of soul, an irradiance of mind, a deeper insight, a broader hope.'[41]

Like Ruskin, Jefferies insists on the spirituality of matter and, therefore, the shallowness of science that deals only in rational calculations based on material facts. Praying to nature, he feels intimations of life beyond the grasp of intellect, remarking: 'The sun was stronger than science; the hills more than philosophy.' Jefferies' insistence on the holism of truth, incorporating all human faculties, parallels his insistence on the oneness of humanity with nature. He writes:

> Why this clod of earth I hold in my hand? Why this water which drops sparkling from my fingers dipped in the brook? ... Matter is beyond understanding, mysterious, impenetrable ... How strange, that condition of mind which cannot accept anything but the earth, the sea, the tangible universe! ... The air, the sunlight, the night, all that surrounds me seems crowded with inexpressible powers, with the influence of Souls, or existences, so that I walk in the midst of immortal things.[42]

Jefferies' language mixes mystical metaphysics with the hard sensory facts of natural environments, presenting the two as a unity instead of an opposition. From this perspective, it is rationalist, artificial 'science' and 'philosophy' that seem unreal when compared with the 'stronger', authentic phenomena of 'sun' and 'hills'. He sometimes signals his anti-rationalist orientation by replicating the mythical imaginative patterns of antiquity. During one of his outdoor reveries, he recalls, 'The disk of the sun rose over the hill, fluctuating with throbs of light; his chest heaved in fervour of brilliance.'[43] This anthropomorphic depiction of the sun evoked the anthropologist Edward Burnett Tylor's recent classification of 'animism', the tendency of 'primitive' societies to figure natural phenomena as living agents.[44] Yet, unlike Tylor, who sought to describe such mythical thinking from without, Jefferies embraces animism as part of his holistic, enchanted vision of nature. At the same time, Jefferies' mention of nature's 'inexpressibility' exemplifies another pervasive rhetorical pattern in Victorian literature about nature: the impossibility of representing it.

Unwritable Nature and the Search for Authenticity

Alongside the poetics of revelation, Victorian authors frequently deprecated their efforts to represent nature. This strategy extended Gilbert

[41] *Ibid.*, p. 10. [42] *Ibid.*, pp. 20, 41–2. [43] *Ibid.*, p. 15.

[44] Edward Burnett Tylor, *Primitive Culture*, 2 vols. (London: John Murray, 1871), I, pp. 377–453.

White's earlier efforts to move beyond dry taxonomy to depict the dynamic aspects of wildlife in the field (see Chapter 1), an uncontrolled space where systematic surveys blurred with leisurely enjoyment and aesthetic appreciation. Idealist, reverent natural history constructed itself around the desire for authentic experience of nature in contrast to the supposedly second-hand artifice of materialist science. This desire for authenticity both reflected and reacted against consumerist culture, which promised to extend to the masses the pleasures elites enjoyed. Alison Byerly argues that Victorian literary realism was symbiotic with a discourse of 'virtual travel' whereby various genres and media promised to transport audiences far afield by generating 'an almost physical sense of presence'.[45] Popular natural history similarly offered readers a substitute for the exotic specimens that filled the cabinets and menageries of rich collectors. At the same time, the idealist orientation stressed the vanity of such collections, nature's true value being immaterial and spiritual. This conflicted attitude harmonised with a rambling form of narrative, inherited from travel writing, characterised by an 'aesthetics of curiosity'. Such narrative roved restlessly from one object to the next in an endless chain of tantalising novelties.[46] More factual science writing in the period presented itself as an organised representation of nature, describing ideal types of rock, plant or animal, one by one, as though directing the reader around an imaginary museum. Conversely, the nature ramble proceeded as a succession of impressions, not so much guiding the reader through nature as getting lost in it.

Ruskin helped to shape this wandering aesthetic for Victorians by stressing the uniqueness of natural objects from an aesthetic perspective in contrast to the quantifications of plain science writing. The calculative tenor of modernity, he suggested, mechanised perception, blinding people to the reality of nature's infinite variety. As he wrote in the first volume of *Modern Painters*: 'There is no bush on the face of the globe exactly like another bush . . . nor two leaves on the same tree which could not be told one from the other.' Ruskin insisted that an 'ideal form' did exist behind nature's 'mass of various, yet agreeing beauty' but such ideality could be glimpsed only indirectly through the 'imagination' and never within

[45] Alison Byerly, *Are We There Yet? Virtual Travel and Victorian Realism* (Ann Arbor: University of Michigan Press, 2013), p. 5.

[46] Nigel Leask, *Curiosity and the Aesthetics of Travel Writing, 1770–1840* (Oxford: Oxford University Press, 2002), pp. 15–53. See also the American John Davidson Godman's *Rambles of a Naturalist* (1833) as an early example of the genre.

physical nature itself.[47] George P. Landow thus observes that Ruskin viewed art as a way of meditating on nature rather than imitating it, since nature's variety resisted imitation.[48] For Ruskin, to attempt to describe and image nature comprehensively, as modern science sought to do, was to produce an impoverished parody of it. As in Carlyle's distinction between 'manufacture' and 'creation', nature's sublime unrepresentability testified to its greatness.

The natural history populariser Margaret Gatty exemplified the wandering aesthetic, and the accompanying sense of nature's sublime unwritability, in her book *Waifs and Strays of Natural History* (1871). The book's chapters, she explained, 'are arranged in no systematic order, nor upon any definite plan. They come before the reader as "Waifs and Strays" from the ocean come ashore – drifted or floated at the pleasure of winds and waves'.[49] Nature's resistance to representation is particularly pronounced in the book's last chapter, entitled 'A Rambling Essay'. The segment relates her journey to the Isles of Scilly as a fantastical 'dream' of strange impressions. Her description wavers between factual reportage and the tropes of a romance tale, framing 'rambling' as a mode of both physical exploration and inner meditation. Noting the 'jutting rocks' around the island, she ponders, 'But were they rocks indeed, or castles and towers, or what had once been such, lying now under some magic spell? – so fantastically did those stern granite formations rise from the sea, that one might well suppose it – some with turret tops, other pinnacled – white sea-birds settling on them from time to time, motionless as sentinels on guard.'[50] The dreamlike character of Gatty's experience derives from her holistic vision combining rational knowledge of nature's facts with imaginative suggestion and religious feeling. The islands' gorgeous exotic plants reflecting the summer sunshine, she states, produced 'a radiancy of splendour which it was difficult to look at without emotion'. Gatty recalls feeling alienated from her companions, who discussed the facts of the islands' wildlife with careful precision. 'I was not scientific', she writes, driven by 'stronger feelings' than zoological pedantry. As with Ruskin, Gatty's language accentuates the impossibility of capturing nature's infinite variety in any artificial representation. Having noted that 'words' are 'but the signs of things experienced or known of', she describes the 'leaves' of tropical

47 Ruskin, *Works*, III, pp. 145–6.
48 Landow, *Aesthetic and Critical Theories of John Ruskin*, pp. 49–52.
49 Margaret Gatty, *Waifs and Strays of Natural History* (London: Bell and Daldy, 1871), p. 1.
50 *Ibid.*, pp. 150–1.

plants on the islands, 'if leaves they may be called, ribless and veinless as they were'. To list these gorgeous plants, she remarks, 'were but to name names'.[51] Nature's appearances, Gatty implies, surpass humans' abilities to verbalise them, so that depicting her explorations as a 'dream' perhaps comes closer to the truth than any scientific description could do.

The rambling mode, combined with an insistence on nature's unwritability, could encourage a proto-ecological sense of nature's interconnectedness across both time and space. Gosse can be seen to use 'rambling' forms of narrative throughout his career to highlight the holistic irreducibility of nature's forms. His first book, for example, proceeded as a dialogue between a father and son as they explored natural environments in Canada. Justifying this conceit, Gosse wrote,

> [B]y taking the reader, as it were, and transporting him into the midst of the very scenes and objects represented, a life and a vigour might be preserved, which would be wanting in a formal narrative. And many little trifles might be thus touched, which . . . all help to make up a true picture. Thus, too, we may ramble from one subject to another (as the humming-bird waywardly shoots from flower to flower), often by a transition more abrupt than could be permitted in a systematic discourse.[52]

Gosse's humming-bird simile registers his identification with the wildlife that he studies, and his feeling of being immersed in it rather than observing it from outside. The form of his book accentuates the artifice of 'systematic' discourses isolating life-forms that, in reality, live always in relation to each other. Gosse's *Rambles on the Devonshire Coast* pushes this point further by narrating his wanderings as an endless chain of connections. Each object he encounters leads to other, contiguous or otherwise related objects. Remarking the high cliffs behind him leads into a description of 'the Jackdaws, which were playing there', and this digression further involves a kestrel and gull that come to spar with the jackdaws for dominance of the territory. Noticing the pebbles of the beach, he recalls, 'It struck me that I might find something under them,' and, turning them over, he slowly reveals a crowd of colourful sea-worms and molluscs. Having prised these creatures from their stony dwellings, Gosse examines them at home and finds them diminished to 'little shapeless lumps of fibrous jelly'. However, when he immerses them in a vase of seawater, 'all of the clustering branchiae are separated and waved, and the

[51] *Ibid.*, pp. 160, 152, 156.
[52] Philip Henry Gosse, *The Canadian Naturalist* (London: John Van Voorst, 1840), p. viii.

long oral tentacles are thrown from side to side, and the pellucid animal glides quickly along with a graceful even motion'.[53]

These vignettes illustrate the organisms' interdependence with their geological, chemical and biotic environments. Gosse suggests that abstracting organisms from their living contexts effaces much of their existence. In this way, the natural history 'ramble' represented not only a convenient device for engaging readers but also an argument for the value of field studies. Field and museum closet had competed for authority in the life sciences since the eighteenth century when Linnaeus had approached organisms as isolated specimens to be classified while Buffon had sought to document organisms' lives in the wild.[54] Closet and (and later laboratory) workers tended to denigrate field studies for their comparative lack of control and precision.[55] Gosse defended the rambling approach explicitly in another memoir:

> Natural History is far too much a science of dead things; a necrology. It is mainly conversant with dry skins . . . with objects . . . impaled on pins, and arranged in rows in cork drawers . . . [T]hat alone is worthy to be called Natural History, which investigates and records the condition of living things . . . their affections and passions . . . their various arts and devices, to protect their progeny, to procure food, to escape from their enemies, to defend themselves from attacks; . . . the countries in which they dwell; their connexion with the inanimate world around them.[56]

Gosse's rhetoric echoes White's interest in wild animals' 'conversations' (see Chapter 1). It also recalls Carlyle's description of history as 'a thousand blending movements' and 'an ever-living, ever-working Chaos of Being, wherein shape after shape bodies itself forth from innumerable elements'.[57] Similarly as Carlyle conceived of history as an infinite web of relations from which the historian could never fully abstract himself, Gosse depicts natural history as an immersive experience, not a body of objects to be scrutinised from without.

The rambling form serves similar ends in Jefferies' writings. The preface to his memoir *Wild Life in a Southern County* (1879) echoes Gosse's

[53] Gosse, *Rambles*, pp. 8, 11–12.

[54] See Ann Shelby Blum, *Picturing Nature: American Nineteenth-Century Zoological Illustration* (Princeton, NJ: Princeton University Press, 1993), pp. 318–19.

[55] See Dorindra Outram, 'New Spaces in Natural History', in Ronald Rainger, Nicholas Jardine, James A. Secord and Emma C. Spary (eds.), *Cultures of Natural History* (Cambridge: Cambridge University Press, 1996), pp. 249–65.

[56] Philip Henry Gosse, *A Naturalist's Sojourn in Jamaica* (London: Longman, Brown, Green and Longmans, 1851), pp. v–vii.

[57] Thomas Carlyle, 'Thoughts on History', *Fraser's Magazine*, 2 (1830), 413–18 (p. 414).

argument that nature's interconnectedness demands narrative representation that is digressive and impressionistic. 'Nature is not cut and dried to hand', Jefferies writes, 'nor easily classified, each subject shading gradually into another.' For example, the life of a starling 'cannot be separated from the farmhouse in the thatch of which it often breeds, the rooks with whom it associates, or the friendly sheep upon whose backs it sometimes rides'.[58] Jefferies' listing of the bird's acquaintances shows the poverty of closet studies in terms of not only their narrowness, but also their atemporality. The starling's existence can only be fully understood through protracted observation of its behaviour in the field. At the same time, Jefferies suggests that the imagination must be allowed to ramble to glimpse the historical aspect of nature in which the present is determined by a vast concatenation of previous states and events. In *The Story of My Heart*, his reveries in a valley lead him to conceive of the present as connected with the ancient past similarly as one field connects with another. Drinking in the landscape, Jefferies recalls, 'I saw back through space to the old time of tree-ferns, of the lizard flying through the air, the lizard-dragon wallowing in sea foam, the mountainous creatures, twice-elephantine, feeding on land.'[59] The landscape Jefferies perceives is haunted by many millennia of change.

Also, like Ruskin, Jefferies emphasises nature's sublimity by denying the possibility of verbalising it. Defending the obscurity of his discourse, he asserts: 'Clumsy indeed are all words the moment the wooden stage of commonplace life is left.' His theatrical image associates words with artificial contrivance, alienated from cosmic noumena. He deprecates his description as 'rude stones of thought', linking language not so much with nature as with the primitive culture described by Tylor, with humankind's (still) rudimentary efforts to master the universe.[60] In one of his last essays, Jefferies all but dismissed writing nature as futile, since the first-hand impressions nature made were too subtle and personal to find expression in common language. Considering the changing tints of a dandelion in the sun, he despaired: 'There are a million books, and yet with all their aid I cannot tell you the colour of the May dandelion . . . So many, many books, and such a very, very little bit of nature in them!'[61] Such statements amounted to a stylistic pose, similar to 'realist' orientations of the period, in which Jefferies attacked conventional representations of nature in order to position his work as somehow closer to it. While asserting the futility of

[58] Richard Jefferies, *Wild Life in a Southern Country* (London: Smith, Elder and Company, 1879), pp. vii–viii.
[59] Jefferies, *Story of My Heart*, p. 17. [60] *Ibid.*, pp. 3, 32.
[61] Richard Jefferies, *Field and Hedgerow* (London: Longmans, Green and Company, 1889), p. 23.

efforts to capture nature on the page, he nonetheless sought to trace nature's contours by emphasising its elusiveness.

This technique also appears in William Henry Hudson's late Victorian narratives of wildlife observation. In *Nature in Downland* (1900), he praises Keats' poetry for its vivid fidelity to nature's aesthetic-emotional power. Yet Hudson complains that anyone who feels that such writing expresses 'as much as he has felt in nature . . . does not inhabit the same world with me'.[62] Language fails to convey direct experience of nature for Hudson because it is communal and thus impersonal – a bridge between people rather than an authentic expression of their inner lives. Hudson imagines a universal experience of nature that runs deeper than artificial language, but this experience defies articulation. The writer can only urge readers to see nature for themselves. In *Birds in a Village* (1893), he comments, 'Who delights not in a bird? Yet how few among us find any pleasure in reading of them in natural history books! The living bird, viewed closely and fearless of our presence, is so much more to the mind than all that is written.'[63] Compared to direct experience of the natural world, literature seems a Platonic shadow. Indeed, in an effort to identify himself more closely with authentic nature, Hudson poses (somewhat unconvincingly) as an outsider to literature. Despite his erudition, Hudson declares: 'I am not a great reader, and know few books.'[64] He had, he mused, 'perhaps' been 'too long a doer, herding too much with narrow foreheads, drinking too deeply of the sweet and bitter cup, to experience that pure unfailing delight in literature which some have'. Such delight, he suggests, is proportionate to the enervation of 'the natural man' inside us.[65] Hudson's view of literature as a form of alienation from nature recalls Ruskin's complaint that modern life was afflicted by 'the disease of thinking', which obstructed direct experience.[66]

These examples show that framing nature as revelatory could render nature writing a contradiction in terms since language and the self-consciousness associated with it estranged humanity from nature. The rhetoric of nature's unwritability distances such texts from the 'ecomimesis' critiqued by Timothy Morton, which seeks to hide its own representational ontology.[67] Instead, these texts exhibit the ecopoetic 'negativity'

[62] William Henry Hudson, *Nature in Downland*, 2nd edition (London: Longmans, Green and Company, 1900), p. 166.

[63] William Henry Hudson, *Birds in a Village* (London: Chapman and Hall, 1893), p. 124.

[64] Hudson, *Downland*, p. 7. [65] Hudson, *Birds*, p. 180. [66] Ruskin, *Works*, XVI, p. 126.

[67] Timothy Morton, *Ecology without Nature* (Cambridge, MA: Harvard University Press, 2009). See Chapter 1.

advocated by Kate Rigby that foregrounds the 'inadequacy of literature as a vehicle of representation or mode or response'.[68] Such representational negativity fits with the epistemic humility of conservative idealism, which stressed nature's suprahumanism. The Victorian texts discussed here thus prefigure the 'restorative ruses' of recent British nature writing that seeks to re-enchant landscapes (see Chapter 4).

Nature Appreciation and Social Class

The construction of nature as a scene of elusive beauty and mystery deeper than science could fathom or literature could represent fitted with an elitist sensibility that deplored incipient mass culture and consumerism. Authors often framed nature studies as a vehicle of aesthetic refinement, which placed one's taste above that of the vulgar mob. Recall Ruskin deploring the vulgarity of materialist Darwinism. Similarly, Kingsley asserted that the true naturalist valued each object 'not carnally, as the vulgar do, by its size or its pleasantness to the senses, but spiritually, by the amount of Divine thought revealed to him therein'.[69] In the same vein, an 1845 primer on natural history had contrasted 'the careless lounger' who marked only the vague 'symptoms of life around him' with 'the curious and enlightened wanderer' whose 'prying eye and mind intelligent not only can appreciate the grosser beauties of the scene, ... but perceive objects of wonder multiply at every step he takes'.[70] Valuing wild environments in this way as stores of inimitable beauty and potential spiritual transcendence functioned as a means of social distinction, displaying one's cultural superiority to utilitarian philistines. In this way, conservative idealism in Victorian nature writing could replicate the class hierarchy often implicit in the pastoral tradition. Whereas the pastoral had opposed the aristocrat taking pleasure in natural landscapes to the (often hidden) peasants labouring in them, the authors discussed here would seem to oppose cultivated appreciation of nature from more vulgar interactions with it, either as work or leisure. Yet such cultivated appreciation could also be imagined as arising out of physical labour and, therefore, attainable regardless of social status.

68 Kate Rigby, '"Come Forth into the Light of Things": Material Spirit As Negative Ecopoetics', in George C. Stallings, Manuel Assensi and Carl Good (eds.), *Material Spirit: Religion and Literature Intranscendent* (New York: Fordham University Press, 2014), pp. 111–28 (p. 117).

69 Kingsley, *Glaucus*, p. 45.

70 Thomas Rymer Jones, *The Natural History of Animals*, 2 vols. (London: John Van Voorst, 1845–52), I, p. 1.

Kingsley seems at first to associate nature studies with the leisured elite by describing them as a preferable alternative to upper-class country sports. The best naturalist, he declares, may also be 'the sportsman' who knows natural environments well and has the free time to explore them. For the angler, particularly, Kingsley offers studies of the river and its inhabitants as a cure for the boredom of hours without a bite. Imagining the reader in this situation giving up, lighting a cigar and snoozing on the bank, Kingsley demands: 'Now how much better, instead of falling asleep, to have walked quietly round the lake side, and asked of your own brains and of nature the question, "How did this lake come here? What does it mean?"'[71] Kingsley's countryside appears a playground for the rich to pursue more or less edifying pleasures. The human labour that maintains this playground is largely absent.

At the same time, however, he also presents nature studies as a way of reforming leisure, making it energetic and productive so that it becomes another kind of work. Peter Bailey notes that constructions of 'rational recreation' reflected the progress of bourgeois values, which invested labour with a new nobility and deprecated aristocratic leisure as lazily effete.[72] Kingsley thus stresses the physical challenge of nature studies, identifying it with a middle-class masculinity based on strength, industriousness and independence. Natural history, he writes, consists in 'active, self-helping life'. Far from being, 'effeminate or pedantic', he asserts, 'our perfect naturalist should be strong in body; able to haul a dredge, climb a rock, turn a boulder, walk all day, uncertain whether he shall eat or rest . . . he should know how to swim for his life, pull an oar, sail a boat, and ride the first horse which comes to hand.'[73] By presenting naturalists like himself as muscular doers rather than effeminate epicureans, Kingsley downplays the differences between the labouring and leisure classes. The two further blur in his discussion of the relations between hunting and natural history. After remarking on the useful knowledge of 'the sportsman' for studying nature, Kingsley comments that 'one has often longed for a scientific gamekeeper or huntsman' to become an author in this line. He conflates the experiences of wealthy riders and shooters with those of the paid professionals who serve them. Further, he addresses himself not to a purely leisured elite, but to a broad audience with varied employments and leisure opportunities. Nature studies, he writes, is a passion shared by

[71] Kingsley, *Glaucus*, pp. 16, 19.
[72] See Peter Bailey, *Leisure and Class in Victorian England* (London: Routledge and Kegan Paul, 1978).
[73] Kingsley, *Glaucus*, p. 43.

'the cultivated man . . . pent up in the drudgery of London work' and the 'portly head-clerk' who spends 'many a pound . . . out of his small salary' on 'glazed and corked drawers full of delicate insects' he has collected.[74] Kingsley's construction of nature studies as leisurely work open to all ameliorates the sharp divisions between labourer and leisured appreciator implicit in the pastoral. Instead, nature studies internalises the work of 'improvement' celebrated in the georgic. As agriculturists cultivate the land, naturalists cultivate themselves. Kingsley envisages this process dissolving class divisions, declaring breezily: 'The successful investigator of the minutest animals takes place unquestioned among men of genius, and is considered, by virtue of his science, fit company for dukes and princes.'[75] The cultural capital associated with reverent nature studies would seem open to all.

The line between labour and leisure is further blurred in Miller's *Old Red Sandstone*, which recounts his discovery of geology through his work as a humble quarryman. Miller recalls his younger self awaiting the first day of his apprenticeship with dread. Following a carefree youth, the young Miller thought, 'now I was going to exchange all of my day-dreams, and all of my amusements, for the kind of life in which men toil every day that they may be enabled to eat, and eat every day that they may be enabled to toil!'[76] However, he finds life in the quarry less tedious than expected, thanks to fossils embedded in the rock, which stir his imagination. Puzzling over how diverse forms came to share the same strata, Miller reports, 'I was lost in conjecture, and found I had food enough for thought that evening, without once thinking of the unhappiness of a life of labour.' The quarry, he reflects, showed him 'that a very exquisite pleasure may be a very cheap one, and that the busiest employments may afford leisure enough to enjoy it'. Miller describes his experience as confirmation of William Cowper's vision of labour as 'the primal curse . . . soften'd into mercy'.[77] The quotation highlights the influence of the georgic in Miller's construction of natural history as work, and in the sense not of drudgery but of ennoblement. In line with Victorian medical prescriptions of 'recreation', Miller's geology constitutes less a rest from his labours than a supplement to them, exercising the mental faculties left unstimulated by his manual tasks. Like Kingsley, Miller suggests that by working upon nature (even as a leisure activity), humans improve themselves, similarly as the work of agriculture improves the land. This orientation tends to smooth over class tensions, levelling the poor and wealthy. Diarmid

[74] *Ibid.*, pp. 5, 17. [75] *Ibid.*, pp. 6–7. [76] Miller, *Sandstone*, p. 4. [77] *Ibid.*, pp. 6–8, 3.

Finnegan notes that Victorian natural history societies embodied 'middle-class values', and Miller illustrates this point by opening *The Old Red Sandstone* with a denunciation of Chartist activism. Rather, he advises, working men in 'cultivating your minds' so that they might match the upper classes 'in intelligence', 'power' and 'happiness'.[78]

This construction of nature appreciation as a form of personal improvement somewhere between physical work and leisurely contemplation may account for Hudson's seemingly contradictory claims to being at once a cultivated aesthete and a rough 'doer' apart from the world of literature. Contrary to antitheses of bodily labour and mental culture, Hudson suggests that the two are inseparable and modernity's blindness to this reality is the root of its philistinism towards both nature and art. Whereas Miller's quarrying shifts into exaltations at the fossil record, Hudson's aesthetic appreciation of landscapes and animals merges with the work of observing them at length. This mingling of labour and leisure echoed Ruskin's idealisation of medieval crafts as activities that combined the discipline of work with the imaginative, spiritual stimulation of art. However, although Hudson's appreciations of the landscapes include their agricultural labourers, he remains at a remove from these people, treating them as aesthetic objects. *Nature in Downland* recalls an encounter with a young Sussex shepherd, who strikes Hudson as 'a very pretty picture' amidst the sky and grass. He elides his conversation with the youth, focusing on the boy's visual traits such as his 'perfectly Saxon' face, 'which you see more in Sussex than anywhere in England'.[79] This depiction frames the shepherd-lad as another of the landscape's native fauna, subject to Hudson's gaze but unable to return it. As in John Berger's critique of the zoological gaze and Laura Mulvey's of the male one, the Sussex shepherd is denied subjectivity, reduced to an object of Hudson's pleasure.[80] In such moments, the optimistic, georgic model of social mobility through the merging of labour and leisure falters. The labourers' closeness to the landscape renders them one with it while the naturalist-aesthete appreciates both from a higher vantage point.

Jefferies further underlined the gap between landscape appreciation and labour. Focusing on agricultural workers, he suggested that their economic precariousness rendered their experiences of the landscape very different from those of the leisurely observer. Recalling one particularly

78 *Ibid.*, pp. 2–3. 79 Hudson, *Downland*, pp. 46–7.
80 See Jonathan Berger, *Ways of Seeing* (London: Penguin, 1972), p. 47; Laura Mulvey, 'Visual Pleasure and Narrative Cinema', *Screen*, 16 (1975), 6–18.

hard season, Jefferies writes of corn reapers working themselves thin and haggard amidst a scene that would have looked idyllic to the rich idler. Endless toil robs the workers of the time and energy to enjoy the land they know better than anyone. Jefferies comments on the workers 'bitter' fate:

> Here was the wheat, the beauty of which I strive in vain to tell you, in the midst of the flowery summer, scourging them with the knot of necessity; that which should give life pulling the life out of them, rendering their existence below that of the cattle, so far as the pleasure of living goes ... This is human life, real human life – no rest, no calm enjoyment of the scene ... the hard fist of necessity forever battering man to a shapeless and hopeless fall.[81]

Jefferies was no socialist, but his vision of agricultural drudgery diverges sharply from Miller's picture of manual labour transitioning easily into aesthetic reverie. Instead, he depicts field work as painful, monotonous degradation. His rendering of it as a repetition of humanity's 'fall' from Eden contradicts Victorian celebrations of work as the engine of human improvement. Jefferies also dissents from idealisations of rural labour in opposition to industrialism, which, Ruskin complained, turned men into machines. Jefferies' farm workers fare no better than this, their tough working conditions 'pulling the life out of them' and 'battering' them out of shape as though they were livestock being pummelled into meat. In contrast to Kingsley and Miller's georgic optimism, he laments the absence of pastoral leisure in most human lives, whether under the feudal past or capitalist modernity.

These divergent views prefigure uncertainties about the overlap or gulf between labour and leisure among twentieth-century authors discussed in the next chapter. The cultivated nature appreciation associated with conservative idealism was not straightforwardly pastoral or georgic: it hovered ambiguously between work and play, between practical activity and personal enrichment. It simultaneously embodied liberalism's ethic of self-improvement and rejected its culture of mass consumerism. Further, as the next section shows, like Rigby's negative ecopoetics, the logic of cultivated nature appreciation could energise an ethics of animal welfare and environmental protection.

Caring for Nature

The latter decades of the nineteenth century witnessed an upsurge in campaigning against cruelty to animals, particularly through the anti-vivisection

[81] Jefferies, *Field and Hedgerow*, p. 133.

movement. At the same time, the growth of shooting into a mass pastime jeopardised wild populations as never before. New organisations of naturalists such as the British Ornithologists' Union (BOU), founded in 1858, promoted the study of Britain's wildlife and came to spearhead early efforts to protect such populations. The Cambridge zoologist Alfred Newton, who helped to found the BOU, was instrumental to the first parliamentary legislation to limit bird hunting in 1869. The growth of the idea that wild species should be shielded from anthropogenic extinction was indexed by the establishment of the Society for the Protection of Birds in 1889 (it would not receive the 'Royal' appellation until 1903). What exactly, though, made wild species intrinsically valuable? Newton had appealed to human utility, arguing that birds helped sailors to navigate British coasts.[82] However, a more powerful impetus for conservation came from efforts to preserve aesthetically appealing landscapes from industrial development such as in the Lake District Defence Society and National Trust for Places of Historic Interest and Natural Beauty. This logic can be traced through several writers' celebrations of natural environments and their inhabitants in these decades. The model of nature studies as a refined, moral pleasure also encouraged some authors to promote sympathy for wild animals and to condemn cruelty towards them. John Parham argues that the era witnessed 'an early environmental activism and "green politics"' founded upon 'a broadly materialistic awareness' that humans subsisted in a network of relations with other species. However, he hedges that this awareness was 'incomplete' as Victorians generally demurred from conceiving of biodiversity as an intrinsic good in need of protection.[83] Their discourse remained generally anthropocentric.

Ruskin's view of nature as a scene of moral-spiritual types drove him to revile field sports since the destruction of wild species would deprive future generations of stimuli to the conscience. He clarifies this point in *Love's Meinie* (1873–81) by arguing that avian forms have long formed models for the celestial and demonic in art. Focusing on the swallow, he writes: 'Type sometimes of the stranger, she has softened us to hospitality; type always of the suppliant, she has enchanted us to mercy; and in her feeble presence, the cowardice, or the wrath, of sacrilege has changed into the fidelities of sanctuary.' Ruskin sees the craze for hunting such

[82] Henry M. Cowles, 'A Victorian Extinction: Alfred Newton and the Evolution of Animal Protection', *British Journal for the History of Science*, 4 (2013), 695–714.

[83] John Parham, 'Introduction', *Green Letters*, 14 (2011), 5–9 (p. 5).

creatures as indexical of humans' spiritual decline. He laments that, for the modern leisure class, 'the principal business of its life is the killing of sparrows . . . [N]o English gentleman in recent times has ever thought of birds except as flying targets, or flavourous dishes.' His singling out of the sparrow indicates the godlessness of modernity, alluding to the biblical dictum that God decides the fate of every sparrow, and, in Ruskin's interpretation, 'would not see [it] perish without pity' (see Thomas Bewick's similar remarks on sparrows, discussed in Chapter 1).[84] His references to 'targets' and 'flavourous dishes' further evokes animal sensuality. The nation's inability to appreciate birds as living presences dovetails with their moral degradation, moved only by base aggression and appetite. Ruskin's valorisation of wild nature against a society that values it only as something to be consumed fits with his wider critique of political economy. His dictum 'There is no wealth but life' could incorporate the nonhuman living world as a factor in humans' spiritual well-being.[85]

Jefferies' aesthetic-spiritual view of wildlife sometimes similarly leads him to encourage readers to lower their guns and raise their binoculars. He exemplifies this tendency near the end of *The Amateur Poacher* when recalling an afternoon spent stalking a pheasant. Finally getting the creature in his crosshairs, Jefferies writes:

> My finger felt the trigger . . . but in the act I hesitated, dropped the barrel, and watched the beautiful bird. That watching so often stayed the shot that at last it grew to be a habit: the mere simple pleasure of seeing birds and animals, when they were quite unconscious that they were being observed being too great to be spoiled by the discharge.[86]

Jefferies suggests that observing wild animals at length opens the mind to a higher pleasure than killing them. Like the vulgar pleasures of food or sex, killing offers gratification that is immediate but also fleeting. By contrast, the pleasure of watching animals alive gratifies endlessly. Like curiosity, its satisfaction inflames further desire in an infinite loop. Jefferies highlights the nobility of this pleasure by depicting it as a sudden intuition that flashes upon him like a religious epiphany. Like Ruskin's privileged seer, he perceives a value in wildlife to which he was previously blind.

Jefferies occasionally made explicit appeals for the protection of wild species. In *The Life of the Fields* (1884), he ends a lovingly detailed description of the wild otter with a plea for protection of this creature

[84] Ruskin, *Works*, XXV, pp. 71, 17–19. See Matthew 10:29. [85] *Ibid.*, XVII, p. 105.

[86] Richard Jefferies, *The Amateur Poacher* (London: Smith, Elder and Company, 1879), p. 237.

that could then still be found swimming in the Thames. Jefferies comments:

> The truth is the otter is a most interesting animal and worth preservation, even at the cost of what he eats . . . This animal is one of the few perfectly wild creatures that have survived without any protection from the ancient forest days . . . It would be imagined that its occurrence so near the metropolis would be recorded with pride; instead of which, no sooner is the existence of an otter suspected than gun and trap are eagerly employed for its destruction. I cannot but think that the people of London at large, if aware of these facts, would disapprove of the attempt to exterminate one of the most remarkable members of their fauna. They should look upon the inhabitants of the river as peculiarly their own.[87]

Jefferies' rhetoric frames Britain's otters as national treasures, as worthy of protection as any building (see Chapter 4). His rhetoric of shared communal property evokes ramblers' campaigns for the 'right to roam' across private rural land. Conceiving of rural Britain as national property pushed back against centuries of enclosures that had driven the masses into towns and cities and turned much of the countryside into a playground for elites. Like Ruskin, Jefferies bases his conservationism on non-utilitarian value. The otter is 'interesting', he asserts, not only due to its appearance and behaviour, but also as a link with England's past, descending from 'forest days' more ancient than any dynasty. For Jefferies, the otter's capacity to rouse the imagination trumps the economic 'cost' of protecting it. Urbanites allow its destruction, he implies, out of ignorance, unaware of the otter's less tangible value. Jefferies' rhetoric illustrates the links between nature aesthetics, distaste for vulgar consumerism and late Victorian conservationism. He considers wild species as valuable not in their own right but as aids to humans' inner cultivation.

The model of nature studies as moral cultivation could also lead Victorians to imagine interspecies sympathy as intrinsic to natural history. This tendency can be seen in William Warde Fowler's *A Year with the Birds* (1886), which relates his bird-watching experiences. Fowler asserts that, to distinguish different species on the wing, the ornithologist must make 'a personal acquaintance – a friendship, as I must call it in my own case – with these little birds, as they live their every-day life among us'. For Fowler, understanding wild animals requires caring for them and imagining life from their perspective in a way that resonates with later concepts of empathy. 'You may shoot and dissect them', he comments, 'but a bird is

[87] Richard Jefferies, *The Life of the Fields* (London: Chatto and Windus, 1884), pp. 124–5.

a living thing, and you will never really know him till you fully understand *how* he lives'.[88] As Victorian historians identified national temperaments among peoples, Fowler, a historian by training, claims that visually indistinct species can be told apart by their human-like '*character*'. Hence, the sedge-warbler 'is a restless, noisy, impudent little creature, not at all modest or retiring, and much given to mocking the voices of other birds'. Conversely, the reed-warbler 'is quieter and gentler', sitting and crooning on a single branch while its cousin 'like a fidgety and ill-trained child, is never in one place, or in the same vein of song, for more than a minute at a time'.[89] Fowler resembles an indulgent schoolmaster, chiding the naughty sedge-warbler with barely concealed affection. Unlike the dry, quantitative facts observed by the closet naturalist, the field wanderer's sum of observations represents a more humanistic knowledge in which birds are not only *known about* but *known* in the sense of personal, sympathetic relations.

Interspecies sympathy or empathy features even more strongly in Eliza Brightwen's *Wild Nature Won by Kindness* (1890), which details the habits of wild animals she nursed as unorthodox 'pets'. Initially publishing portions of the book in *The Animal World*, the periodical of the Royal Society for the Prevention of Cruelty to Animals (RSPCA), Brightwen presented the study of animals as an ethical and, therefore, sentimental activity. Like Warde Fowler, she describes her subjects as 'friends', explaining: 'I use the word friendships advisedly, because truly to know and enjoy the society of a pet creature you must make it feel that you are, or wish to be, its friend, one to whom it can always look for food, shelter, and solace; it must be at ease and at home with you before its instincts and curious ways will be shown'. Brightwen frames sympathy with wild animals as a prerequisite to gaining knowledge of them, since a creature in fear of its keeper will not display any other aspects of its character. She further points out that animals unhappy in their captivity are likely to soon perish and so yield no new insights into their kind. Without heartfelt understanding of its wants, she writes, 'the poor little captive will possibly be miserable, pining for liberty, hating its prison, dreading the visits of its jailor, and so harassed in its terror that in some cases the poor little heart is broken, and in a few hours death is the result'.[90] Brightwen's animals resemble the slaves of abolitionist literature, inviting readers to imaginatively project

88 William Warde Fowler, *A Year with the Birds* (London: Macmillan and Company, 1891), p. 44.
89 *Ibid.*, pp. 80–1.
90 Eliza Brightwen, *Wild Nature Won by Kindness* (London: T. Fisher Unwin, 1909), p. 12.

themselves into the same position.[91] This analogy was common in RSPCA literature, which derived from the same evangelical culture as abolitionism. Brightwen's claimed emotional insights into her captives legitimise her presence in the male-dominated literature of natural history, restricting her authority to the traditionally feminised sphere of non-rational feeling.[92] Her habit of bringing animals into her home and 'taming' them with 'kindness' further associates her nature studies with feminine domesticity. Instructing readers in how to rear chicks away from their nests, she declares: 'if the weather is cold they should be near a fire, as they miss the warmth of the mother bird, especially at night'.[93] Brightwen's claim to understand the chicks' longing presents her studies as a kind of mothering. Such conservationism and humanitarianism were limited, though, by the anthropocentrism inherent in the model of nature studies as a form of self-cultivation. Ruskin and others' criticisms of the destruction and persecution of wildlife usually framed these trends as undesirable because they were morally degrading *to humans*. Natural environments came to be imagined as precious because they were sources of aesthetic (and, therefore, moral) stimulation. By centring human interests, such discourse could sometimes be reconciled with anthropogenic extinction, environmental destruction and animal suffering if these outcomes were perceived as necessary to human progress.

A strong example of this compromised environmentalism can be seen in Kingsley's work. Although he raged against the unsanitary conditions of modern cities and hailed rural tradition for its supposed healthy naturalness, Kingsley also depicted human 'improvement' of nature as part of God's plan. This georgic thinking is exemplified in his essay 'The Fens' (1867), which celebrates the draining of the East of England's coastal plain for agriculture. The wild landscape that preceded this action was no Eden, he asserts, but 'riot and chaos'. Unregulated by human intelligence, peat moss and beaver populations gradually sank 'stately forests' into a 'Dismal Swamp'. Curiously, Kingsley presents such unguided natural processes as parallels to the chaotic workings of laissez-faire capitalism. He writes that, by felling trees and damming streams, the beavers resembled 'selfish

91 On abolitionism and sympathy or empathy, see Marcus Wood, *Slavery, Empathy, and Pornography* (Oxford: Oxford University Press, 2003).

92 See Carolyn Merchant, *The Death of Nature: Women, Ecology and the Scientific Revolution* (San Francisco, CA: Harper and Row, 1980); Val Plumwood, *Feminism and the Mastery of Nature* (New York: Routledge, 1993).

93 Brightwen, *Kindness*, p. 19.

speculators in all ages, sacrificing freely the public interest to their own'.[94] The opposition for Kingsley is not authentic nature versus artificial society but wise management around a central authority versus anarchic, atomised self-interest, both in nature and society. The disappearance of old wild species is thus nothing to mourn, he argues, since their very existence was symptomatic of disorder, created by the Fall. Such fauna, he writes, are 'now happily extinct in the fens . . . For instead of mammoth and urus, stag and goat, that fen feeds cattle many times more numerous than all the wild venison of the primaeval jungle; and produces crops capable of nourishing a hundred times as many human beings.' While he admits to feeling 'A certain sadness' when he imagines the loss of so many varied and beautiful species, Kingsley insists that 'Reason and conscience tell us, that it is right and good' to turn 'a waste and howling wilderness' into 'a garden of the Lord'.[95] Civilisation is not the enemy of nature, Kingsley implies, but the realisation of its destiny. Nature unguided by wise authority seems as dangerous and unhealthy as society unguided by wise authority. In either case, human interests and self-actualisation reign supreme.

Ruskin went further than Kingsley in his condemnation of industrial civilisation, opposing activities inimical to wild habitats such as the building of railways and artificial reservoirs. Mark Frost contends that Ruskin's lecture 'The Work of Iron' (1858) expressed 'nascent ecological thinking' as Ruskin illustrated humanity's co-dependence with wider natural processes; in this case, iron oxidisation.[96] By rusting, Ruskin observes, iron fertilises the soil, enabling plants and crops to grow. Iron in its naturally occurring state thus fits into wider 'circles of vitality' that humans cannot live without.[97] Nonetheless, Ruskin's environmental vision remained anthropocentric, generally valuing the natural world as a vehicle for humans' development. In *The Queen of the Air* (1869), he worries that industrialisation alienates people from cultural and religious heritages, which are produced by sympathetic engagement with nature. If, as he declares, myths 'first arose among a people who dwelt under stainless skies . . . we certainly cannot read their story, if we have never seen anything above us in the day but smoke, nor anything around us in the night but candles'.[98] Ruskin suggests that alienation from nature alienates humans from art and all creative works of the spirit, which arise from acquaintance

[94] Charles Kingsley, 'The Fens', in *Prose Idylls, New and Old* (London: Macmillan and Company, 1880), pp. 93–131 (pp. 111, 113).
[95] *Ibid.*, pp. 93, 119–20. [96] Frost, 'John Ruskin, Environment', p. 21.
[97] Ruskin, *Works*, XVI, p. 378. [98] *Ibid.*, XIX, pp. 301–2.

with the natural world. Nature must be preserved not for its own sake but to prevent humans from turning into unfeeling, unimaginative machines.

This premise also underlies Ruskin's protests against bird-shooting in *Love's Meinie*. He often seems more concerned with what birds have meant to the human soul through the ages than with their welfare in the present. 'The mythology of the bird', he declares, is 'the most important part of its natural history.' The destruction of such wildlife is immoral, Ruskin implies, because doing so is to destroy the imaginative source of the stories through which humans make sense of themselves. He reinforces the linkage between natural landscapes and imaginative life by envisaging an England ruled utterly by the logic of political economy: 'a country without angels and without birds, without prayers and without songs, without trees and without flowers'.[99] Ruskin's elision of the natural world with religious culture underlines his belief that the latter depends on experience of the former. Ruskin did not care to know of the ruthless natural economy revealed by Darwin because he was only interested in nature as an instrument for improving human happiness (which he saw as bound up with moral and spiritual development). 'To watch the corn grow, and the blossoms set', he wrote, 'to draw hard breath over ploughshare or spade; to read, to think, to love, to hope, to pray, – these are the things that make men happy'.[100] Aspects of nature that did not contribute to this end of happiness were to be avoided, as Ruskin showed in his criticism of anatomy for reducing beautiful animals to gruesome cadavers.[101]

Poetics of Estrangement

Such discourse of mentally connecting with nature or its creatures was opposed in the period by a tendency to construct natural landscapes and nonhumans as radically alien. This tendency derived from paradigm shifts in theories of the Earth's history and a growing conviction that the Wordsworthian sense of nature being specially adapted to the human mind was erroneous.[102] A new consensus had emerged that the Earth was far older than previously thought, forcing theologians to reinterpret the biblical narrative of creation in figurative terms. Human history shrank to a brief moment at the end of a vast timeline of change. Further, the

99 *Ibid.*, XXV, pp. 40, 44. 100 *Ibid.*, V, p. 382.

101 See Dinah Birch, '"That Ghastly Work": Ruskin, Animals and Anatomy', *Worldviews*, 4 (2000), 131–45. Compare Rousseau's comments on anatomy, discussed in Chapter 1.

102 See Susan Lorsch, *Where Nature Ends: Literary Responses to the Designification of Landscape* (Madison, NJ: Fairleigh Dickinson University Press, 1983).

natural-theological approach to the organic world as a system of adaptations tended to construct species as ends in themselves instead of vehicles of human pleasure or guidance. Heidi Scott notes that in the period the model of nature as a balance was countered by an alternative model of instability and chaos.[103] The latter paradigm was enhanced by Darwin's theory of evolution by natural selection, which characterised the world by struggle, random variation and environmental contingency. The mechanisation of observation via laboratory instruments emerged symbiotically with an ethics of 'objectivity', which sought to generate knowledge undistorted by human biases. Simultaneously, secular philosophers and anthropologists emphasised the tendency of humans' self-projections to shape their world views. From the 1850s, researchers began to use the theological term 'anthropomorphism' to critique the projection of human emotions and rationality onto animals.[104] These developments were often matched in Victorian nature writing by a poetics of estrangement that foregrounded nature's strangeness and opacity to the human observer.

An early example of this estrangement can be seen in Hugh Miller's *Scenes and Legends of Northern Scotland* (1835), one passage of which muses on the waves of species extinctions revealed by geology. Miller registers how these discoveries disrupt traditional notions of the world as a stable, benevolent design revolving around humans. Instead, he admits, one might imagine humans as another species that could disappear from existence as prehistoric creatures did. A time may come

> when, like the fossils of these beds, our whole species shall be mingled with the soil; and when, though the sun may look down in his strength on our pleasant dwellings and our green fields, there shall be silence in all our borders, and desolation in all our gates . . . Surely it is well to believe that a widely different destiny awaits us . . . that our very bones, instead of being left, like the exuviae around us, to form the rocks and clays of a future world, shall, like those in the valley of vision, be again clothed with muscle and sinew.[105]

The passage shows Miller disregarding nature as a text of divine intentions. His hope for the second coming derives not from nature but revelation; indeed, his faith is kindled in spite of nature's signs, not because of them.

[103] Heidi C. M. Scott, *Chaos and Cosmos: Literary Roots of Modern Ecology in the British Nineteenth Century* (University Park: Pennsylvania State University Press, 2014).

[104] Paul White, 'The Experimental Animal in Victorian Britain', in Lorraine Daston and Gregg Mitman (eds.), *Thinking with Animals: New Perspectives on Anthropomorphism* (New York: Columbia University Press, 2005), pp. 59–82.

[105] Hugh Miller, *Scenes and Legends of Northern Scotland* (Edinburgh: A. and C. Black, 1835), pp. 57–8.

Miller's image of flesh and bone turning into 'rocks and clays' points to the danger of a purely physical explanation of the world stripping human life of significance. Instead of being served by nature, as in Ruskin's celebration of 'the work of iron', humans risk being absorbed into it, mere material for geological formations.

Such sentiments can be found in much Victorian popular natural history that constructed nature as a worthy study for the faithful but not a basis for faith.[106] Gosse similarly stressed the limits of nature's religious usefulness. He warned that although nature indicated some kind of designer, 'When an anxious conscience demands to know something more of God . . . whether there is forgiveness with Him, and mercy, – the creatures are mute. One says, It is not in me! and another says, It is not in me! All are ominously dumb on such questions.'[107] Gosse's allusion to the book of Job, in which land and sea deny possessing God's wisdom, reinforces humans' reliance on the Bible in a world that seems often ruthless and inhumane.[108] Some authors argued late into the century that divine wisdom and morals could still be detected in the natural order, even via an evolutionary paradigm.[109] However, many others redefined religion as a matter of inner, mental experience, separate from the blind workings of matter.[110]

This idea of nature being irrelevant to human spirituality sometimes finds its way, surprisingly, into Jefferies' pantheist reveries. Although, in *The Story of my Heart*, his transcendental experiences derive from encounters with natural objects and landscapes, Jefferies also stresses that these are only vehicles for accessing something immaterial. 'All nature', he asserts, 'all the universe that we can see, is absolutely indifferent to us, and except to us human life is of no more value than grass. If the entire human race perished at this hour, what difference would it make to the earth? What would the earth care? As much as for the extinct dodo.' Far from feeling solidarity with nature, Jefferies describes it as an antagonist, a 'distinctly anti-human' power to be transcended. The design that humans used to see

106 See Fyfe, *Science and Salvation*, p. 7; Jonathan Topham, 'Natural Theology and the Sciences', in Peter Harrison (ed.), *The Cambridge Companion to Science and Religion* (Cambridge: Cambridge University Press, 2010), pp. 59–79 (p. 60).

107 Philip Henry Gosse, *The Aquarium: An Unveiling of the Wonders of the Deep Sea* (London: John Van Voorst, 1854), p. 210.

108 Job 28:14.

109 See, for example, Arabella Buckley, *Life and Her Children* (London: Edward Stanford, 1880).

110 See Pietro Corsi, *Science and Religion: Baden Powell and the Anglican Debate, 1800–1860* (Cambridge: Cambridge University Press, 1988), pp. 201–3.

in nature, he suggests, turned out to be self-projections imposing order and meaning on a disordered, meaningless world. Thus, he exclaims:

> How extraordinary, strange, and incomprehensible are the creatures captured out of the depths of the sea! . . . They shock the mind because they exhibit an absence of design . . . They have no shape, form, grace, or purpose; they call up a vague sense of chaos, chaos which the mind revolts from . . . But it is the same in reality with the creatures of the earth . . . [Consider] the toad. At its shapeless shape appearing in an unexpected corner many people start and exclaim . . . All the designless, formless chaos of chance-directed matter squats there embodied in the pathway . . . By no course of reasoning, however tortuous, can nature and the universe be fitted to the mind . . . I am separate altogether from these designless things.[111]

Jefferies' opposition between mindful humanity and mindless nature prefigures the words of a speaker in one of Oscar Wilde's later dialogues that 'What art reveals to us is Nature's lack of design . . . Nothing is more evident than that Nature hates Mind.'[112] Like Wilde's aesthete, Jefferies finds nature disappointing if not revolting. If it inspires, it does so negatively, prompting the mind to differentiate itself from such chaos. From Jefferies' perspective, humans' capacity to design, imagine and interpret places them at odds with the physical universe. This point is accentuated by the detail of the 'pathway', a purposeful construction, unlike the product of organic chance that wanders disturbingly onto it. The value that humans see in nature, Jefferies implies, really lies within them.

A similar sense of alienation is discernible in Leslie Stephen's writings on the natural world, which are as much about writing as about nature. In one essay, Stephen confessed that, although he loved the country, 'I love it best in books. In real life I have remarked that it is frequently damp and rheumatic, and most hated by those who know it best.' Writers such as White, he remarked, were 'weavers' of a 'delightful magic'. They offered in their pages an 'ideal mode of rustication' that would always be superior to first-hand experiences of rural landscapes.[113] When Stephen recounted his own experiences of wild nature in his book about mountaineering in the Alps, *The Playground of Europe* (1871), the first two chapters were dominated by literary responses to the region. This orientation enabled Stephen

111 Jefferies, *Story of My Heart*, pp. 57–8, 62. Compare the discussion of Mary Shelley's *The Last Man* in Chapter 1.

112 Oscar Wilde, 'The Decay of Lying', in *The Complete Works of Oscar Wilde*, ed. Ian Small, 8 vols. (Oxford: Oxford University Press, 2000–17), IV, pp. 72–103 (p. 73).

113 Leslie Stephen, 'Rambles among Books: No. I. – Country Books', *Cornhill Magazine*, 42 (1880), 662–79 (pp. 662–3).

to relativise the beauty that his contemporaries found in the mountains, noting that this sensibility had only appeared relatively recently. Before the mid eighteenth century, he claimed, writers were more likely to view a mountain as 'a hideous excrescence'. The beauty of the mountains was not intrinsic, he suggested, but dependent upon their contrast with the 'artificial', homogenous metropolitan society from which the middle-class mountaineer escaped. He reinforced this point through an anecdote of his Swiss mountain guide looking out on London from a train window and pronouncing it 'finer' than the view from Mont Blanc. Stephen went on to quote with relish his favourite poets of the mountains, entertaining the proto-Wildean thought that appreciation of natural landscapes is displaced appreciation of art. 'To my taste', he declared ironically, 'Shelley's poetry is in the most complete harmony with the scenery of the higher Alps; and I think it highly creditable to the mountains that they should agree so admirably with the most poetical of poets'.[114] For Stephen, the pleasure of the Alps lies as much, if not more, in tracing the sources of literary inspiration as in experiencing such inspiration oneself. In contrast to Hudson's pose of being a natural, unliterary man, Stephen suggests that there is no escaping one's cultural shell; indeed, that it is this shell of culture that galvanises nature appreciation.

In Jefferies' writing, such alienation from nature in general was matched by a sense of animals' alterity as mental subjects. While convinced that many nonhumans possessed emotions and reason, Jefferies also dwelt on how their experiential worlds might differ radically from human ones. His efforts to imagine what post-humanists, following the lead of biosemiotician Jakob von Uexküll, would call animals' *Umwelten* require estrangement from one's habitual human perspective.[115] He argues that millennia of attacks from humans must have caused many wild animals to fear the human form, and their fear to fixate on humans' arms. 'Animals, birds, and fishes', Jefferies writes, 'have learned to watch and to dread man's arms. He raises his arms, and in an instant there shoots forth a bright flash of flame, and before the swift wings can beat the air again the partridge is dashed to the ground.' Further, from a nonhuman perspective, Jefferies reasons, these limbs must seem shockingly bizarre. 'Suppose yourself with your arms tucked to your sides under the fur of an animal', Jefferies suggests. 'Something of the idea may be gathered by putting on a cloak

[114] Leslie Stephen, *The Playground of Europe* (London: Longmans, Green and Company, 1894), pp. 21, 1, 63.

[115] See Alberto Romele, 'Towards a Posthuman Hermeneutics', *Journal of Posthuman Studies*, 3 (2019), 45–59.

without sleeves or armholes. At once it will be apparent how helpless all creatures are in comparison with man.'[116] Empathising with nonhumans involves making human forms and perceptions strange. At the same time, Jefferies' admiration of animals' apparent *joie de vivre* can be tinged by the regretful thought that humans are incapable of such immersive pleasure due to their consciousness of their mortality. While humans live in endless fear of death, he claims, animals appear 'so trustful and so content with their fate, resting in themselves and unappalled. If but by reason and will I could reach the godlike calm and courage of what we so thoughtlessly call the timid turtle-dove, I should lead a nearly perfect life.'[117] His meditation points to the impossibility of his enjoying life in the moment as animals seem to do, since 'reason' necessarily abstracts him from life's immediacy. Paradoxically, sympathising with animals can foreground their remoteness.

Similarly, Hudson vacillated between anthropomorphic interpretation of animal motives and critical reflections on the inaccessibility and probable alterity of animal experience. Observing a rook cawing loudly, he reads the noise as 'expressing the old truth that life was worth living', a 'feeling' that he judges 'very common' in the living world.[118] Birds' calls of alarm seem to signify through a natural, cross-species sound symbolism akin to human cries and other vocalisations. Hence, Hudson declares, agitated birds use 'harsh grating or piercing sounds that properly express violent emotions of a painful kind'. Yet, simultaneously, he notes that some birds appear to have lost touch with this primordial sound symbolism and express their agitation in 'mere lispings compared with those emitted by the bird when singing'.[119] The apparent inaptness of such birds' vocalisations undermines the sense of interspecies understanding. Perhaps sounds that strike Hudson as innately expressive of discomfort or fear do not strike other species so. Hudson's interpretation of pleasure in the rooks' caw is commensurable with such thinking since the interpretation depends not on the sound's tonal qualities but its context. The rooks, he remarks, were 'feeding in the meadow grass in that June weather, with the hot sun shining on their glossy coats', far away from wintry 'days of want', 'the anxious breeding period' and 'the annual slaughter of the young birds'.[120] The rooks' apparent joy is more inferred via knowledge of their living conditions than sympathetically felt.

[116] Jefferies, *Life of the Fields*, p. 168. [117] *Ibid.*, p. 53. [118] Hudson, *Birds in a Village*, p. 19.
[119] *Ibid.*, p. 13. [120] *Ibid.*, p. 19.

Whereas Jefferies mourned humans' remoteness from the immediacy of animal experience, Hudson mourns humans' inability to even guess what animals' experiences might be like. Considering insects' apparent similarities to humans in their social organisation, he laments that

> in spite of all our prying into nature's secrets, all our progress and the vast accumulations of knowledge at our disposal, we do not and never can know what an insect knows or feel what it feels . . . Nay, more, we know that these small bodies have windows and avenues which ours are without; that they are conscious of vibrations which for us do not exist . . . We can gaze through a magnifying glass at certain of their complex organs of sense, but cannot conjecture their use.[121]

Hudson's awareness of animals' alterity leads him to suggest that naturalist authors like him occupy a middle ground between artistic sympathy and scientific scepticism. The nature writer at once infers deep motivations behind animals' actions and avoids prying too deeply into these motivations, contenting himself with observation of outward behaviour. Discussing authors' different attitudes to people, Hudson states that he is 'not satisfied with the mere semblance or appearance of things, seeing men as trees and rocks, or as works of art'; yet neither is he 'teased . . . with that restless desire to pry into and minutely examine the secret colour and texture of the mind of every person I meet'. Hudson defines this approach as 'the mental attitude of the naturalist . . . who does not wish to worry his brains overmuch, and likes to see very many things with vision a little clearer than the ordinary, rather than to see a very few things with preternatural clearness and miss all the rest'.[122] Hudson's anti-anthropomorphic natural history substitutes the ecological breadth of animals' interrelated behaviours for the emotional depth of poetry. At the same time, his levelling of wildlife and people as parallel fauna of the Sussex landscape evokes the relationship between place and identity, and the interest of Victorian nature writers like Hudson in defining it.

Geographies of Identity

Roger Ebbatson observes that, as British colonial possessions grew through the nineteenth century, efforts intensified to define Englishness through literary appreciations of England's landscapes. Ebbatson finds in such writing a complex mixture of assurance and self-doubt as expanding

[121] Hudson, *Downland*, pp. 69–70. [122] *Ibid.*, p. 104.

territories prompted the question of where to locate the imperial home or 'centre'. Images of a unitary environment that bound people both in the present and across generations jostled with visions of fragmentation and homelessness.[123] This ambivalence permeated Victorian nature writing, which both reflected and helped to shape discourses of nativeness and foreignness. Natural history as a mass pastime was catalysed by the development of the railways, which enabled people of various classes to explore rural landscapes on short trips with authors such as Gatty and Gosse as their guides. Similarly as travel writing had long constructed the nation as a succession of vignettes, nature writers characterised Britain as a collection of separate locales simultaneously united in distinction from other countries. Their depictions of Britain's 'natural' environments could also blur with urban ones, stereotypically associated with cosmopolitanism and ethnic rootlessness. At the same time, the model of nature as a set of discrete, national territories and their indigenous life forms could be undermined by visions of migration as a defining trait of the living world, which reached its apex in global trade, exploration and empire. The latter tendency could relativise nativeness, reframing organic environments as scenes of constant displacement and renewal via incoming species.

This tension between local differences and national oneness was reflected in depictions of variations in the flora and fauna across regions. 'It would be as unreasonable to look for Devonshire myrtles in Yorkshire gardens as for Devonshire algae in its waters', Gatty wrote, stressing the importance of climate, sea saltiness and coastal terrain in the maintenance of different species. Yet, at the same time, she made this point in her book *British Sea-Weeds* (1862), a title that implied a unity beneath the apparent diversity. Even as Gatty's descriptions of sea life revealed its tendency to transgress borders, she still supported the dichotomy of native and foreign through her designations of different species. Hence, although she wrote that 'Portuguese men-of-war' could be 'found along the coasts of Cornwall and Devonshire', these creatures remained foreigners, displaced from their native 'hot latitudes' by the Gulf Stream along with other 'warm-sea-loving plants'.[124] Gatty's anthropomorphic description of the hydrozoans evokes Victorian racial typologies, by which different collective temperaments were held to match different ancestral homelands. In his

[123] Roger Ebbatson, *An Imaginary England: Nation, Landscape and Literature, 1840 –1920* (Aldershot: Ashgate, 2005), pp. 1–8.

[124] Margaret Gatty, *British Sea-Weeds*, 2 vols. (London: Bell and Daldy, 1872), I, p. xvi.

History of Civilization in England (1857), Henry Buckle had argued that climate, soil, food and forms of landscape shaped different peoples' mental propensities, rendering character the product of environment. Gatty's sea creatures echo this divisional logic, cleaving to their natural habitats and perishing if they stray away from them.

The conflict between the ideology of national unity and the reality of geographical variation in Britain's wildlife further emerges in Kingsley's *Glaucus*. Although his narrative focuses on the Devonshire coast, Kingsley treats it as a microcosm of the nation. Surveying the creatures washed up on the beach, he invokes different names for them from around the British Isles, noting 'The great red star-fish, which Ulster children call "the bad man's hands"; and the great whelks, which the youth of Musselburgh know as roaring buckies'. Like Ruskin, Kingsley switches between Latin nomenclature and homely folk names to balance international science against the experiences of local communities around Britain. He identifies a sea cucumber first as 'Cucumaria Pentactes', before stating: 'but he has many a stout cousin round the Scotch coast, who . . . submits, among the northern fishermen, to the rather rude and undeserved name of sea-puddings'.[125] His depiction of these varieties of sea creatures as cousins parallels the supposed national affinity of the English and Scots, beneath superficial differences. Studying Britain's wildlife thus becomes a way of discovering and reaffirming its people's shared historical experience. The mythical and artistic resonances Ruskin saw in nature become nationalised, turning coastlines into repositories of folk memory. Although Kingsley viewed the English (and, by extension, lowland Scots) as Teutons who had conquered the Celts, he accepted that Celtic elements endured in Britain's language and history, producing a composite nation (although not an equal one).[126] He sees Britain's environments as mirroring this complex history as successive waves of incoming species have overwhelmed older pedigrees. Hence, Kingsley examines coral specimens 'from that curious "Rat Island"' in the Bristol Channel 'where still lingers the black long-tailed English rat, exterminated everywhere else by his sturdier brown cousin of the Hanoverian dynasty'.[127] Britain's natural history seems to echo the mythical clash of races, indicating divine endorsement of Saxon supremacy.

125 Kingsley, *Glaucus*, pp. 64, 116.

126 See Stanwood S. Walker, '"Backwards and Backwards Ever": Charles Kingsley's Racial-Historical Allegory and the Liberal Anglican Revisioning of Britain', *Nineteenth-Century Literature*, 62 (2007), 339–79.

127 Kingsley, *Glaucus*, pp. 121–2.

The notion that Britain's natural landscapes were saturated with ancestral memories emerges more strongly in Jefferies' writing. Late Victorian biology witnessed debates over the possibility of racial memory – that is, ancestral experiences persisting in the imaginative intuitions of later generations.[128] Jefferies sometimes suggests that Britain's wild environments can draw out these inherited experiences. In *The Open Air* (1885), he writes:

> Under the trees the imagination plays unchecked, and calls up the past as if yew bow and broad arrow were still in the hunter's hands . . . Sit down on the root of this oak . . . and on that very spot it is quite possible a knight fresh home from the Crusades may have rested and feasted his eyes on the lovely green glades of his own unsurpassed England. The oak was there then, young and strong; it is here now, ancient, but sturdy . . . The swine snort, and the mast and leaves rustle as they thrust them aside. So little is changed: these are the same sounds and the same movements, just as in the olden time.[129]

The continuity of (some of) Britain's landscapes makes them seem like vehicles for reconnecting sympathetically with one's ancient predecessors. Jefferies further frames natural landscape as linking individual and national memory in another essay in which a child is imagined speaking to ears of wheat. The wheat tells him of how it witnessed King William II being killed by an arrow in the New Forest, and, before that, the Vikings sailing up England's rivers in longboats. This national history flows into individual ones in Jefferies' phantasmagoria as the wheat states that Vikings came 'all along the river up to the place where the stream in the ditch runs in; just where your pap took you in the punt, and you got the waterlilies, the white ones'.[130] Jefferies' talking vegetation depicts England's wild environments as repositories of memory that anchor individual lives to a collective, national heritage. At times, wild creatures appear to do the same, as Jefferies writes in another essay: 'I own I love the blackbird . . . There is a bold English independence about him – an insolent consciousness of his own beauty. He must somehow have read Shakespeare, for he seems quite aware of his "orange-tawny bill" and deep black hue.'[131] Jefferies' fanciful vision of the blackbird swelling with pride at Shakespeare's recognition of him frames English wildlife and culture as deeply intertwined. Each, he suggests, derives its power partly from the other.

[128] Laura Otis, *Organic Memory: History and the Body in the Late Nineteenth and Early Twentieth Centuries* (Lincoln: University of Nebraska Press, 1994).

[129] Richard Jefferies, *The Open Air* (London: Chatto and Windus, 1885), pp. 185–6.

[130] *Ibid.*, p. 12. [131] Jefferies, *Wild Life in a Southern Country*, pp. 164–5.

Peter Coates notes that the fetishising of 'certain landforms, places, and creatures' as 'native nature' in late nineteenth-century North America correlated with growing stigmatisation of incoming flora and fauna as foreign. This modern fear of invasive species, he suggests, emerged in tandem with anxieties about mass immigration harming the nation.[132] A similar parallel can be observed in some late Victorian nature writing as concerns about migration merged with the notion that city life was degenerate. Jefferies, for example, noted the preponderance of foreign trees and vegetation on London's outskirts, complaining,

> The foreigners oust the English altogether. Let the cedar and the laurel, and the whole host of invading evergreens, be put aside by themselves, in a separate and detached shrubbery, maintained for the purpose of exhibiting strange growths. Let them not crowd the lovely English trees out of the place. Planes are much planted now, with ill effect . . . [I]t is unsightly. It has no association. No one has seen a plane in a hedgerow, or a wood, or a copse. There are no fragments of English history clinging to it as there are to the oak . . . The hawthorn is a part of natural English life – country life. It stands side by side with the Englishman, as the palm tree is pictured side by side with the Arab . . . Most of the foreign shrubs about these semi-country seats look exactly like the stiff and painted little wooden trees that are sold for children's toys, and, like the toys, are the same colour all the year round.[133]

Jefferies' distaste for the foreign plants revolves less around their individual features than their incongruity within England's wider environment and history, both natural and cultural. Their ugliness is contextual rather than intrinsic, deriving from their displacement from their 'native' surroundings. Jefferies associates the incongruity with a general urban malaise in which people value only wealth and become estranged from the authentic heritage of the countryside. His comparison of 'foreign shrubs' in England to painted toys frames them as artificial commodities alienated from organic processes. In his apocalyptic novel *After London* (1885), Jefferies envisioned the city collapsing into a toxic swamp, inimical to all life. His criticism of London's 'strange growths' traces such urban degeneration to the city's alienation from nature in which all forms subsist in their proper places. Jefferies presents his preference for native flora in England as not just personal whim, but a universal law inscribed in nature. England's native animals can be seen to agree with him, he claims. 'Birds . . . are attracted by our own trees and shrubs. Oaks are favourites with rooks and

[132] Peter Coates, *American Perceptions of Immigrant and Invasive Species: Strangers on the Land* (Berkeley: University of California Press, 2006), pp. 3–4.

[133] Richard Jefferies, *Nature near London* (London: Chatto and Windus, 1887), pp. 189–203.

wood-pigeons; blackbirds whistle in them in spring . . . Thrushes are fond of the ash, and sing much on its boughs.'[134] Nature, Jefferies implies, abhors displaced foreigners; as beings that are part of nature, humans should abhor them too.

Yet Victorian writers did not always view Britain's landscapes so ethnocentrically. Hudson's writing complicates the boundaries between the native and foreign, perhaps due to his own complex heritage. Born to American parents of Anglo-Irish descent, Hudson spent his childhood and youth in Argentina before moving to England in his thirties. It is thus tempting to read his discussions of the indigenousness of 'British' birds for echoes of his own insecurity about whether he truly belonged in Britain, a country where he long struggled to attain literary distinction. Hudson begins his book *British Birds* (1894) by appearing to reinforce the nationalisation of wildlife. He admonishes ornithologists who 'regard as "British" any species of which one specimen has been found in a wild state within the limits of the United Kingdom'. Many such birds, he wrote, were 'lost wanderers', not 'citizens' but 'aliens', called British only 'by a pleasing fiction'. Hudson's language would seem to harden the discourse of exclusive identitarianism. However, his quibbling over the definition of avian 'Britishness' also illustrates the category's elasticity. Hudson distinguishes between Britain's winged 'residents', 'regular visitants' and 'Occasional Visitants', replacing the binary of natives and foreigners with a more nuanced spectrum of belonging.[135] Indeed, his designation of birds as 'residents' avoids the nationalist baggage of nativeness, foregrounding everyday habitation rather than some imagined hereditary essence.

Hudson's sense of the flexibility and open-endedness of Britishness as a designation for wildlife is sharpened by his awareness of the changing fortunes of different populations over time. Regarding the greylag goose, he observes that this bird, 'formerly a common summer resident' in England, 'is now confined as a breeding species to a few localities in Scotland and the Hebrides', and even there it is 'rapidly diminishing'. In the future, he predicts, this bird will only 'visit our coasts' as 'a winter migrant from northern Europe'.[136] Hudson's shifting language thus shows that the natural world's unstable conditions (further destabilised by human activity) can turn natives into foreigners. By the same token, perhaps 'foreign' species could find a home in Britain as Hudson with his mixed

[134] *Ibid.*, p. 204. Ebbatson shows a more ambivalent side to Jefferies' representation of national identity in *An Imaginary England*, pp. 69–82.

[135] William Henry Hudson, *British Birds* (London: Longmans, Green and Company, 1895), p. xx.

[136] *Ibid.*, p. 228.

background had striven to do. Noting in *Birds in a Village* (1893) that hunting had wiped out many beautiful species, he suggested importing foreign ones to brighten up the countryside. Against the objection that birds from other lands could not thrive in Britain, he points to examples where they have. For example, the red-legged partridge, despite being 'a stranger' brought over from the Continent, 'has spread throughout the greater part of England'. Even more impressively, he writes, the pheasant, originally from Asia, has adapted to 'our cold, wet climate . . . and [now] practically it is one of our wild birds'.[137] Hudson shows that nativeness is partly illusory as species once regarded as foreign come to seem part of an eternal landscape. The thriving of these 'stranger' species upon British soil problematises the notion of a distinct British natural history existing in isolation from the rest of the world.

Hudson further complicates the pedigree of British nature writing by suggesting that appreciation of Britain's wildlife can be developed through non-British literary influences. Encountering a hill in rural England inhabited by various birds, he wishes 'that I could build a cabin and live there all the summer long, forgetful of the world and its affairs, with no human creature to keep me company, and no book to read'. His proscription of reading is ironic, given that his imagined project for returning to nature strongly follows the American example of Thoreau's *Walden* (1854). Further, Hudson then qualifies his vow to give up all literature, making an exception for the works of the Spanish poet Juan Melendez. Melendez's 'artlessness', he claims, induces a 'feeling of kinship with all the children of life' more than any British poet, even with regard to British wildlife. Indeed, Hudson doubts 'the boast we sometimes make, that the feeling for nature is stronger in our poets than in those of other countries'.[138] Hudson's cosmopolitan literary hinterland undercuts Jefferies' nativist suggestion that British nature can only be interpreted through the lens of British literary tradition. On the contrary, Hudson suggests, outsiders' perspectives are likely to discover fresh resonances in the flora and fauna.

Such troubling of eco-nationalist categories was paralleled by the blurring of boundaries between rural and urban spaces, which, in the pastoral mode, demarcated 'nature' from artificial civilisation. This blurring is exemplified in Brightwen's aforementioned descriptions of animals nursed in her house, although her language still entrenches a dichotomy between external wildness and internal, domestic tameness. The binary is complicated, though, by Brightwen finding these animals in the grounds of her

[137] Hudson, *Birds in a Village*, pp. 94–5. [138] *Ibid.*, p. 29.

mansion home on the edge of London. While this space lies outside of her domestic hearth, it is not quite wild either, being a set of cultivated gardens surrounded by walls. Indeed, Victorian naturalists might not even need to venture out of their houses to discover phenomena worth studying. 'We doubt not', Gosse wrote, 'if any one were to lay himself out patiently to watch, and carefully to observe, the habits of the Mice that infest our cupboards, or the Sparrows that throng our housetops, he would find highly interesting and unknown facts resulting from his minute observations.'[139] Gosse's comment destabilises the nature/culture binary, depicting human homes as simultaneously the habitats of wild animals. Like Brightwen, Gosse often brought his zoological work home with him, building aquariums (a word he coined) so as to reconstruct the ocean and observe its live inhabitants in domestic comfort. However, while his aquariums exemplified georgic control over and improvement of nature, Gosse's image of the mice in the cupboards is more levelling. '[L]ay[ing] himself out' to observe mice poaching kitchen supplies, Gosse's naturalist finds wildlife surviving and adapting itself literally under humans' noses.

Jefferies further collapses the nature/culture divide in an essay on pigeons nesting in the masonry of the British Museum. Similarly as he encouraged readers to look closely at rural landscapes and realise how much they previously missed, Jefferies shows that the elusive beauty of wildlife can also be found in the heart of London. The birds' 'small rounded bodies', he writes, are 'so motionless and neutral-tinged that they might be mistaken for some portion of the carving.' Jefferies makes explicit the point hinted at by Gosse that the subsistence of such wild creatures amidst the trappings of civilisation undermines humans' assumed separateness from and superiority to the natural world. To the pigeons searching for food, Jefferies reflects, 'the building is merely a rock, pierced with convenient caverns . . . The heavy roll of the traffic in Oxford Street, audible here, is nothing to them; the struggle for money does not touch them . . . Nor the many minds searching and re-searching in the great Library.' The image of the human readers scouring through the same books places them in parallel to the birds, 'searching and re-searching' the grounds outside for morsels. Jefferies suggests that, when compared to us with our lives of labouring after elusive happiness, the pigeons enjoy 'the preferable portion'. 'They have not wasted their time', he remarks, 'looking among

[139] Philip Henry Gosse, *An Introduction to Zoology*, 2 vols. (London: Society for Promoting Christian Knowledge, 1844), I, p. x.

empty straw for the grain that is not there.'[140] While Jefferies' commentary leads towards an affirmation of spirituality over material riches and rationalism, the text also questions the opposition of wild nature to civilisation. Undomesticated life intrudes upon the city architecture, turning an artificial structure into another habitat.

Conversely, in another essay, Jefferies contradicts pastoral laments about the destruction of rural nature by industrialisation, claiming that agricultural machinery can become part of the 'natural' landscape. 'The earth has a way of absorbing things that are placed upon it,' he writes. 'The machines sent forth to conquer the soil are conquered by it, and become a part of it, and as natural as the old, old scythe and reaping-hook.' Discussing the threshing machines used for harvesting grain, Jefferies argues that these contraptions 'fit in with trees and hedges, fields and woods', producing 'more varied colours' in the landscape than appeared there before.[141] Jefferies' assertion of the machine's aesthetic harmony with the field forms part of a challenge to the simplistic view that technology alienates humans from the natural world. The machine, he explains, is merely a development of the ancient practice of 'reaping', which 'is all old as the hills'. Contra associations of industrial machinery with death and destruction, Jefferies presents the machine as peacefully coexistent with the field's flora and fauna. He notes that meadow farmers mostly leave the ancient hedges intact due to the shade and shelter they provide to cattle. Consequently, 'hawthorn and blackthorn, ash and willow . . . briar and bramble . . . are still there as in the old, old time'. By focusing on hedges, Jefferies foregrounds the inextricability of human (and specifically English) life from more-than-human nature, since the hedge is at once an artificial structure and an organic growth and habitat for wild organisms. 'Without hedges', he declares, 'England would not be England. Hedges, thick and high, and full of flowers, birds, and living creatures . . . I love their very thorns. You do not know how much there is in the hedges.'[142] Jefferies thus opposes the pastoral view of modernity as a break from nature with a messier, georgic sense of humans existing in various, changing forms of entanglement with the wider living world. He eschews the ideals of both a golden age of harmony with nature and societal progress through domination of it. Rather, he suggests, humanity's relationship with more-than-human nature is a complex, ongoing

[140] Jefferies, *Life of the Fields*, pp. 252–3, 256.
[141] *Ibid.*, pp. 132, 143. For more on Jefferies and technology, see Roger Ebbatson, *Landscape and Literature 1830–1914: Nature, Text, Aura* (Basingstoke: Palgrave, 2013), pp. 115–24.
[142] Jefferies, *Life of the Fields*, pp. 139–40.

negotiation of interests and mutual dependencies. Nonhuman life in England is bound up with humans' material, aesthetic and emotional wants, from the food on their plates to their sense of identity and belonging.

This chapter has shown that Victorian nature writing conceptualised the relationships between nature and nation, and nature and society, in uncertain, conflicting ways. It envisaged identity and being at different times as essentialist and fluid. The British (and particularly English) countryside could be imagined as a discrete community of unique life forms shaped by the country's particular history and geography, which incoming, foreign organisms disturbed and degraded. On the other hand, it could also be conceived of as part of a larger continent-wide or even worldwide organic continuum characterised by mobility and adaptation to different environments. Linked with this ambivalence about nature's (in)congruence with nationalism was uncertainty over where to draw the line between nature and society, and even doubt that such a line could be drawn at all. Jefferies in particular sometimes seems to prefigure Morton's notion of 'ecology without nature', replacing dualist oppositions of human/nonhuman and wildness/civilisation with a logic of complex entanglement and interdependence.[143]

This discussion has highlighted the generic variety and ideological contradictions the category of Victorian nature writing must contain. The writers considered here often channelled comparable intellectual influences to Thoreau, but, with the exception of Jefferies and Hudson, their work tended to be less confessional. Although authors wrote about their feelings and imaginative reveries, they tended to mention their wider personal lives only briefly and elliptically. Science and literature (and, perhaps more crucially, science and religion) were increasingly opposed in the period, but this opposition was also strongly resisted, and much of what we might classify as Victorian nature writing blended these enterprises. The intellectual paradigm of conservative idealism enabled authors to construct empirical study of nature and aesthetic-spiritual experience as complementary. Yet such texts could also be haunted by new scientific ideas that rendered the natural world and its inhabitants alien and estranged from human sympathies. This sense of nature's alterity would intensify in twentieth-century British nature writing. The conservative idealist attitude simultaneously embodied the logic of consumerism and

[143] Timothy Morton, *Ecology without Nature: Rethinking Environmental Aesthetics* (Cambridge, MA: Harvard University Press, 2007).

social mobility and opposed them. It depicted the natural world as a realm of morally improving pleasures that raised its appreciators above the vulgar materialism of urban society, secular science and mass culture. Such nature appreciation hovered ambiguously between the categories of labour and leisure, and the contrastingly aristocratic and liberal values associated with them. Numerous scholars in recent years have identified early examples of ecological or proto-ecological discourse in Victorian literature.[144] Yet, while recognising the potential value of such 'strategic presentism' for rethinking both past and present, this chapter also underlines the inability of most Victorians to imagine nature as intrinsically valuable irrespective of humans.[145] This anthropocentrism is reflected in Victorian nature writers' persistent occupation with identity as they looked to natural environments as vehicles for discovering themselves both as individuals and as members of national or international collectives. Nonetheless, as the final example from Jefferies shows, Victorian nature writing sometimes approaches Lawrence Buell's definition of the 'environmental text' in terms of 'suggest[ing] that human history is implicated in natural history' and conceptualising 'the environment as a process rather than as a constant'.[146] Only in the twentieth century, however, would British nature writers come to articulate an ethics that might be called fully ecological.

[144] See Parkins, *Victorian Sustainability*; John Parham, *The Environmental Tradition in English Literature* (Aldershot: Ashgate, 2002); Hall, *Victorian Ecocriticism*.

[145] Jesse Oak Taylor uses this term in 'Where Is Victorian Ecocriticism?', *Victorian Literature and Culture*, 43 (2015), 877–94.

[146] Lawrence Buell, *The Environmental Imagination: Thoreau, Nature Writing, and the Formation of American Culture* (Cambridge, MA: Belknap Press of Harvard University Press, 1996), pp. 7–8.

CHAPTER 3

Modern

Introduction: Interrogating the Modern

There are many different ways of being modern. Openness to change, adherence to 'progress' and the embrace of technological innovation are all acknowledged to be general characteristics of the modern. But so too are their opposites. 'To be fully modern', Marshall Berman remarks, 'is to be anti-modern': to rebel against the very conditions that one has created for oneself.[1] It is to be uncertain about who one is, what one knows and where one stands; as Berman further suggests, 'to be modern is to live a life of paradox and contradiction' in which it is perfectly possible to be both revolutionary and reactionary, to be excitingly 'alive to new opportunities for experience and adventure' while remaining 'frightened by the nihilistic depths' towards which at least some of these experiences and adventures appear to converge.[2] It is also to be unsure *when* one is, and if being modern is about being cannily attuned to the times in which one lives, it is also about being painfully aware that these times are continually shifting. Berman lists three phases of modernity: the start of the sixteenth century through to the end of the eighteenth century, the revolutionary wave of the 1790s and the fragmented modernity of the twentieth century, 'a modern age that has lost touch with the roots of its own modernity'.[3]

Useful though these historical markers are, it perhaps makes more sense to see modernity in terms of a radical scrambling of temporal coordinates: to be modern is thus simultaneously to be projected into an uncertain future and haunted by an equally indeterminate past. 'Crisis' is an overused term, to be sure, but there is still a key sense in which to be modern is to

[1] Marshall Berman, *All That Is Solid Melts into Air: The Experience of Modernity* (New York: Simon and Schuster, 1982), p. 14.
[2] *Ibid.*, p. 4. [3] *Ibid.*, p. 17.

exist in a more or less permanent state of crisis, in which the hope of a different way of living is dogged by doubts about its viability, or by deeper fears about the viability of *any* way of living: a potentially paralysing form of existential despair. Modern nature writing, as we are seeking to argue in this book, confronts this state of crisis head on, and it would thus be a mistake to see it as fundamentally consolatory or escapist in its inclinations even if modern nature writers, whose work is largely concerned with the here and now, the place-based registering of present experience, have been by no means averse to using their writing to reflect on the imaginative possibilities of other ways of living, other places, other times.

For early to mid twentieth-century nature writers, there was admittedly a lot from which to escape. As the late historian Eric Hobsbawm recalls it, the Short Twentieth Century (1914–91) was an age of extremes marked, especially in its first half, by a concatenation of potentially world-ending catastrophes: an age in which 'the great edifice of nineteenth-century civilization crumpled in the flames of world war'.[4] Twentieth-century British nature writers from Henry Williamson to J. A. Baker were traumatised by war, whether viscerally experienced (by Williamson) or fearfully intuited (by Baker). Some sought solace in their work, taking heart in the regenerative properties of nature even as all was laid waste around them. Others, though, Williamson and Baker among them, saw violence as a constitutive feature of the natural world, a property shared by humans and animals alike, and one inscribed onto landscapes that had long since ceased to be 'natural' in any meaningful sense of the word.

Nature writing in the twentieth century, as will be seen in both this chapter and the next, responded to the crises of the times as well as those broader, existential crises that accrued to an age that seemed to some to have lost its moral bearings, and in which 'British values', upheld to some extent during the epochal wars that threatened them, came to be challenged by regional particularisms, emergent devolution movements and perhaps above all by increasingly tangible, if still frequently disavowed, processes of imperial decline.[5] Like the Romantic and Victorian periods that preceded it, albeit in intensified form, the Modern age presented British nature writers with a triple crisis that seemed to many to be encapsulated in the nation, but whose material and ideational scope reached far beyond it: the

[4] Eric Hobsbawm, *The Age of Extremes, 1914–1991* (London: Abacus, 1995), p. 22.

[5] See, for example, Piers Brendon, *The Decline and Fall of the British Empire* (New York: Vintage, 2008). For a study of the cultural effects of imperial decline as these were felt in Britain, especially England, see also Jed Esty, *A Shrinking Island: Modernism and National Culture* (Princeton, NJ: Princeton University Press, 2003).

environmental crisis, the crisis of representation and the crisis of the modern 'split' self. These crises are writ large in most general descriptive accounts of late nineteenth- and early twentieth-century British cultural modernism, although it would be a further mistake to see twentieth-century British nature writing, with some notable exceptions, as modern*ist* in inspiration; indeed, features of most of the conventional (non-fictional) nature writing from the Modern period are its relative lack of formal experimentation, its predominantly anti-cosmopolitan sensibility and its tendency to shy away from interiority, the dizzyingly perspectivist exploration of human and, in some cases, nonhuman subjective worlds.

This has led to a tendency to regard nature writers of the period as writing in reaction to the modernity of their age, whether this modernity is seen as embedded in particular (technological) ideas of the modern or enshrined in the aesthetic principles of cultural modernism. Certainly, as the twentieth century dawned, it did not immediately appear to offer cultural conditions conducive to the flourishing of nature writing. The early decades of the century, violently punctuated by the First World War, were characterised by the intensification of urbanisation and industrialisation, which some saw as direct threats to the integrity of the English countryside, as well as potentially compromising the collective British ethos for which it came to stand, both during the war itself and more explicitly in its aftermath (in the shape of the 'back-to-the-land' movement of the 1920s and 1930s).[6] At the same time, if no more promisingly, the cultural developments that accompanied these socio-economic processes were frequently associated with an inward turn that sought to focus on subjective perception, the life of the mind and the workings of consciousness. Attention to the externals of the natural world would seem to run counter to both of these dominant, if by no means ubiquitous, cultural impulses. In fact, however, early twentieth-century nature writing can equally well be understood as *demonstrating* these impulses, exhibiting traits that are modern and anti-modern, revolutionising and reactionary, by turns.

One group of British writers, loosely dubbed by Richard Mabey as the 'new naturalists', aligned themselves with scientific developments in the emergent fields of ecology and ethology, advocating the adoption of a properly modern attitude towards the natural world.[7] These conservation-minded writers, among them E. M. Nicholson and Charles Elton,

[6] David Matless, *Landscape and Englishness* (London: Reaktion Books, 2016).

[7] Richard Mabey (ed.), *The Oxford Book of Nature Writing* (Oxford: Oxford University Press, 1995), p. 154. For a historical account of the new naturalists, see also Peter Marran, *The New Naturalists: Half a Century of British Natural History* (London: HarperCollins, 1995).

championed a relationship with nature characterised by respectful restraint and the empirical principles of 'an observational ethic'.[8] While recognising that the conditions of modernity – its accelerated pace of change, its unbridled technological power – had the potential to exacerbate existing threats to natural environments, the new naturalists argued that the solution to the dangers posed by modernity lay in further modernisation, for example, the replacement of private by collective ownership, or new policies of protection and regulation that merged the interests of the individual and the institution in service of a socially progressive, but also environmentally responsible state.[9]

As the cultural geographer David Matless argues, the new naturalists' enthusiastic promotion of scientific expertise, their firm commitment to state regulation and their accompanying ideology of self-restraint were all joined to the task of building a new nation in which modern, orderly landscapes were to be managed according to scientific principles, including the 'controlled creation of new habitats for wildlife'.[10] The new naturalists, by and large, saw themselves as field researchers first and writers only second, and their work was characterised accordingly by an unadorned aesthetic of factual presentation, captured in 'sensible prose' and illustrated with a variety of practical maps, diagrams, tables and charts.[11] Read today, much of this work bears little similarity to what most of us would consider nature writing, and indeed many of the new naturalists were doubtful to the point of being dismissive of those kinds of post-Romantic 'fine writing' that wax lyrical about nature's charms. This led some, such as the ornithologist James Fisher, to reposition foundational figures such as Gilbert White, whom Fisher saw as having been 'elevated to the fountain-head of a tradition of elegant nature writing which has [since] developed into a mockery of the honest language of the *Natural History*'.[12] Instead, Fisher declared, White needed to be reinserted into the tradition where he rightly belonged, that of 'British observers [and] field-workers',[13] a tradition in which respect for nature demanded a rigid separation of fact from fancy: 'Once facts are despised, fancies replace

[8] Caroline Hovanec, *Animal Subjects: Literature, Zoology, and British Modernism* (Cambridge: Cambridge University Press, 2018), p. 77.

[9] David Matless sees these early twentieth-century developments in terms of a 'moral modernity' in which preservationists and other environmentally minded actors looked to subscribe to 'a mechanism of orderly progress driven by planning', and a 'new rural social order' was envisaged whose structures were intended to embody 'moral principles [while] offenders [were] to be cleared out' (*Landscape and Englishness*, pp. 80, 70, 74). For further discussion of the implications of these changes for the cultures of British modernism, and for British rural writing in particular, see the second part of this chapter.

[10] *Ibid.*, p. 308. [11] *Ibid.*, p. 350. [12] Quoted in *ibid.*, p. 351. [13] Quoted in *ibid.*, p. 347.

them; and fancies are poisonous companions to the engagement and appreciation of [the natural world].'[14]

Factual accuracy was the main criterion for such writing, although, as will shortly be seen, this shouldn't be taken to mean that the work produced was shorn of aesthetic considerations, and the recent scholarship of Benjamin Morgan among others helps us to relocate it within, if also to some extent against, the tradition of the nineteenth-century descriptive account.[15] Charles Darwin rather than White is probably best seen here as the key figure, although the new naturalists' view of Darwin was no more uniform than that of more self-consciously literary writers, Virginia Woolf prominent among them, who were significantly influenced by his work. As we will go on to explore in the next section of this chapter, the legacy of Darwin looms particularly large in twentieth-century animal writing, which is best seen as operating at the interface of literature and science, blurring the boundaries not just between literary and scientific forms of nature description, but also between literary and scientific ways of knowing and understanding the world.[16] Twentieth-century British animal writing is not necessarily about British animals, nor is it necessarily about animals at all insofar as such writing has historically used animals as projection screens for human interests and concerns.[17]

However, much of the animal writing that emerged during the Modern period is deeply indebted, whether it owns up to this or not, to contemporaneous developments in the fast-developing field of ethology, and it is signally interested in exploring the particularities of animal behaviour even as it recognises the limits of human knowledge in attempting to understand the detailed physiology of animal bodies – and the intricate psychology of animal minds. At one level, the confrontation of these epistemic limits, a typically modernist concern, seemed to condemn twentieth-century animal writers to rehearsing one historical version or another of 'animal alterity'.[18] At another level, it encouraged them, buoyed by the latest scientific discoveries, to reflect further on a theme already familiar to

[14] Quoted in *ibid.*, p. 351.

[15] See Benjamin Morgan, *The Outward Mind: Materialist Aesthetics in Victorian Science and Literature* (Chicago: University of Chicago Press, 2017). For a detailed discussion of nineteenth-century natural-historical descriptive accounts, see also Chapter 2.

[16] Hovanec, *Animal Subjects*, p. 5.

[17] For a further discussion of what Steve Baker calls the 'symbolic availability' of animals, see Baker's *Picturing the Beast: Animals, Identity, and Representation* (Chicago: University of Illinois Press, 2001); see also Philip Armstrong, *What Animals Mean in the Fiction of Modernity* (London: Routledge, 2008).

[18] The idea of animal alterity is central to Carrie Rohman's study *Stalking the Subject: Modernism and the Animal* (New York: Columbia University Press, 2009).

their nineteenth-century predecessors: the fundamental strangeness of the natural world.

The two writers who come most readily to mind here, D. H. Lawrence and Virginia Woolf, are not usually seen as nature writers today, though recent scholarly attempts to establish the parameters of a 'green modernism' have confirmed that both, albeit in markedly different ways, were deeply interested in the ecological interconnection of living things. They were also interested in the possibilities that discourses of nature offered to construct alternative subjectivities at both the individual and national levels, with the land for Lawrence in particular providing the material foundation on which a new, up-to-the-minute Englishness, born out of but also resistant to modern industrialism, might be formed.[19] Some commentators on Lawrence have seen him as promoting 'an essentially Romantic understanding of animal subjectivity [that] emphasized emotional expression over rational intellect',[20] seeking support for this in his essays, which repeatedly call for humans to get back to their animalistic origins, and in which animality consistently functions as 'a symbol for humans' primitive essence, a more authentic way of being than modern civilization allows'.[21] Lawrence's work is more complex than this, however, and it is perhaps above all in his poetry that his professedly anti-mechanistic position rubs up against a view of animals that implicitly draws on the modern scientific principles it ostensibly disavows. This is also implied in John Alcorn's admittedly dated designation of Lawrence as a 'new naturist' [*sic*]. For Alcorn, 'the new naturism is a world of physical organism, where biology replaces theology as the source both of psychic health and moral authority. The naturist is a child of Darwin; he sees man as part of an animal continuum; he reasserts the importance of instinct as a key to human happiness; he is wary of abstraction.'[22] Insofar

[19] Jeffrey Mathes McCarthy, *Green Modernism: Nature and the English Novel, 1900–1930* (London: Palgrave, 2015).

[20] Hovanec, *Animal Subjects*, p. 118. [21] *Ibid.*, p. 120.

[22] John Alcorn, *The Nature Novel from Hardy to Lawrence* (New York: Columbia University Press, 1977), p. x; also quoted in Bonnie Kime Scott, *In the Hollow of the Wave: Virginia Woolf and Modernist Uses of Nature* (Charlottesville: University of Virginia Press, 2012), pp. 34–5. The classic study of Darwinian influence on literature and the arts is Gillian Beer, *Darwin's Plots: Evolutionary Narrative in Darwin, George Eliot, and Nineteenth-Century Fiction* (Cambridge: Cambridge University Press, 2009). Modern and contemporary literature, similarly, is barely thinkable without Darwin, although by no means all twentieth-century writers agreed with him, Lawrence being probably the most notorious example. On Lawrence's conflicted relationship with Darwin, see Ronald Granofsky, *D. H. Lawrence and Survival: Darwinism in the Fiction of the Transitional Period* (Montreal: McGill-Queens University Press, 2003); see also the discussion of Lawrence's equally ambivalent relationship with 'science' in the third part of this chapter. Unsurprisingly, Darwin's is a massive presence in twentieth- and twenty-first-century nature writing, although as is argued

as he subscribes to this organicist view, Lawrence is – as Berman might put it – modern and anti-modern at the same time, and his work can consequently be seen as both a virulent reaction to modern industrialism and an attempt to forge a new, thoroughly ecological understanding of the human/animal/land nexus which, as for other organicists of the period, looks to the local as a source of inspiration for the national, simultaneously celebrating national Englishness and small-scale social and economic life.[23]

While unlikely to have shared many of Lawrence's views – especially his views on sex – the self-styled 'country writer' H. J. Massingham offers a good example of a popular author of the period whose work provides a robustly organicist view of the English landscape co-grounded in practical agriculturalism and the nationalist trope of 'English Earth'.[24] David Matless rightly sees Massingham as subscribing to a 'counter-modern ecology of holistic understanding' rather than its modernising equivalent, but he also considers him as more than just a nostalgic 'countryphile' turning his back on the future and gazing longingly into a part-remembered, part-imagined past.[25] Similarly, twentieth-century country writing, insofar as it is legitimate to see it as anything like a coherent subgenre, is of a piece with the modernity it often seems so eager to denounce. The 'ecological patriotism'[26] that much of this writing seems to support is less attuned to that oxymoronic concept, 'simple pastoral',[27] than to a revitalised form of georgic which, along with its flexible credo of practical agricultural labour, turns fantasies of rural retreat into productive engagements with the natural world. For Jeffrey Mathes McCarthy, writing of this kind is far from escapist simplification; rather it belongs to a complex machinery of representation adapted to the particular cultural needs of post-war England as well as the new technological demands of a rapidly modernising world.[28]

Anticipatory signs of this sensibility can be seen in the influential work of Edward Thomas, who fell near the end of the First World War, and later examples can be found in writers of the 1930s and 1940s, who were

throughout this book many British nature writers, for a variety of reasons, continue to look for inspiration to Gilbert White.

[23] Matless, *Landscape and Englishness*, p. 176. For more on the local-national binary, see Chapter 2.

[24] *Ibid.*, p. 148. [25] *Ibid.*, p. 141. [26] *Ibid.*, p. 307.

[27] See the distinction between 'simple' and 'complex' pastoral in Lawrence Buell, *The Environmental Imagination: Thoreau, Nature Writing, and the Formation of American Culture* (Cambridge, MA: Harvard University Press, 1995). The literature on pastoral is too voluminous to record here; it suffices to say that nature writing, on both sides of the Atlantic, is indebted, not just to pastoral, but also to different *perceptions* of pastoral at different places in different times. Perhaps less often acknowledged is that both British and American nature writing, albeit in different ways, are similarly indebted to *georgic*: for an extended discussion of this, see the second part of this chapter.

[28] McCarthy, *Green Modernism*, p. 117.

confronted with another, in some ways even more devastating world war. A further complicating factor is that by no means all of these writers were English. While the contemporary English nature writer John Lewis-Stempel in *Where the Poppies Blow*, his prize-winning account of the First World War British soldiers' paeans to the English countryside, breezily announces that 'Britain was universally known [as England] in those far off days', several of the authors he cites would have been unlikely to share his confidence, while more nuanced approaches such as Matless' *Landscape and Englishness* make clear that the conflation of 'Britishness' and 'Englishness' is and always has been ideological work.[29] As Matless suggests, the confusion of 'British' and 'English' has a long history, but it probably made most sense in the high imperial context, where 'England [was] assumed to be heart and head of the British Empire', a sentiment still pervasive in the conservative thinking of our times.[30]

This puts a different spin on the notion of ecological patriotism; it also invites us to think about the regional as well as national differences enshrined in modern British nature writers' work. As was suggested in the two previous chapters of this book, there is a sense in which British nature writing can only be held to be modern insofar as it engages with a world whose boundaries exceed those of Britain. Its localism, however convincingly drawn, may well prove deceptive, and its periodic rather than consistent nationalist sentiments belie the cultural as well as geographical and biological variations it inscribes. John Brannigan has recently made the case for an 'archipelagic modernism' based on 'post-imperial and post-metropolitan forms of community and citizenship': one in which supposedly marginal perspectives turn out to be central, and insularity becomes as much a figure for strangeness as protective familiarity in turbulent times.[31] As will be seen in what follows, the study of modern British nature writing stands greatly to benefit from this archipelagic perspective, which embraces the work of remotely located and/or non-English writers who, with the exceptions of two celebrated Scots, Gavin Maxwell and Nan Shepherd, have rarely been taken into the twentieth-century British nature writing fold. It also clearly profits from being seen through a postcolonial lens that examines the connections between the elitist colonial sensibilities that sometimes drive it and the broadly democratising conservationist sympathies it often professes to uphold.

[29] John Lewis-Stempel, *Where Poppies Blow: The British Soldier, Nature, and the Great War* (London: Weidenfeld and Nicolson, 2017), p. 11.

[30] Matless, *Landscape and Englishness*, p. 38.

[31] John Brannigan, *Archipelagic Modernism: Literature in the Irish and British Isles, 1890–1970* (Edinburgh: Edinburgh University Press, 2015), p. 148.

The intertwined histories of conservation and colonialism are apparent in several different works from the period, from William Henry Hudson's at the turn of the century to Gerald Durrell's in the 1950s. These works remind us, not only that British nature writers are not necessarily English, but also that Englishness, the capacious if at times amorphous idea of English ethnicity, is often fashioned abroad.[32] This reminds us, in turn, that nature writing is often as much about travel between places as about invocations of a particular place, and that it shares several characteristics of a genre to which it has sometimes been opposed, the travel memoir: a genre to which both Hudson's and, later, Durrell's work belongs. Similarly, Durrell's work, in particular, explores both the real and imaginative possibilities of human-animal kinship that are present in much of the animal writing of the period, possibilities embedded within the developing sciences of ecology and ethology and the modern conservationist principles they uphold.

The twentieth century, in this respect, yokes nature writing in all its variant forms to the unfolding story of conservation as both an increasingly professionalised scientific pursuit and, as the century wore on, an increasingly global phenomenon linked to the popularisation of scientific knowledge and media-driven debates on the protectionist measures needed to combat species extinction and environmental decline in an ecologically threatened world. The role of nature writers as science popularisers is of course by no means limited to the Modern period and, along with the broad social and intellectual movements that underpin it, it provides a running theme throughout this book. However, it is arguably in the early part of the twentieth century that the theologically inflected natural history which informed much of the writing of the Victorian period begins to morph into that particular set of characteristics we associate today with the self-consciously modern, predominantly secular genre of British nature writing: that particular, readily identifiable combination of scientific enquiry, environmental concern and autobiographical self-awareness that marks the work of contemporary writers such as Robert Macfarlane and Helen Macdonald, even if 'nature writing' remains, for a good number of them, a frustratingly elusive and, for at least some of them, a scarcely tolerable descriptive term.

At the same time, that period sees little consensus on what nature writing actually is, little interest – with some notable exceptions – in defining its parameters, and for some writers at least, Lawrence prominent among them, little patience with compartmentalising scientific methods as

32 See, for example, Robert J. C. Young, *The Idea of English Ethnicity* (Oxford: Wiley-Blackwell, 2003).

integral to deeper understandings of the workings of the natural world. Lawrence would doubtless have been horrified to hear himself described as a nature writer, and even the new naturalists, hardly a coherent grouping in any case, would probably have been surprised. These definitional dilemmas, which haunt the entire history of British nature writing, are writ particularly large in the Modern period, and this might help to account for the singular eclecticism of the material we present in this chapter, which includes the genres of nature poetry, animal writing and the philosophical essay alongside the descriptive category associated most commonly today with nature writing, a semi-specialised, environmentally conscious form of non-fictional prose.

The period itself is difficult enough to define, and this helps to account in turn, not just for the temporal sweep of this chapter (the early 1900s to the early 1970s), but also for the significant overlaps that occur between it and the next. Henry Williamson, J. A. Baker and Ronald Blythe, for example, all of them writers whose work belongs to any analysis of nature writing in the Modern period, feature in the next chapter rather than this one, although this can partly be explained by the argument made in Chapter 4 to the effect that contemporary nature writing is 'late modern' in style and sensibility: an argument which, as will be seen, also holds for at least some of the material examined in Chapter 3. At any rate, as argued throughout this book, chronology only gets us so far in seeking to better understand the modern history of British nature writing; a genealogical approach seems more suitable in allowing us to identify discursive strands that both cut across particular historical periods and complicate the idea of periodicity itself.

In the rest of this chapter, we follow some of these strands and examine their interconnections, organising our material around the two broad groupings of 'animal writing' (in the first section) and 'rural writing' (in the second). In the 'animal writing' section, we put forward an argument that is further developed in Chapter 4 – namely, that modern animal writing offers a set of philosophical as well as scientific reflections on the entangled human-animal relationship in which the multifunctional figure of the animal other is alternately antithetical to and isomorphic with a divided human self. In the 'rural writing' section, likewise, we seek to show how writing of the period negotiates historical tensions between country and city, hinterland and coast, centre and periphery that effectively give the lie to constructed oppositions between them, using these tensions to interrogate modernism's metropolitan biases while exploring the countryside as an alternative locus of the modern in rapidly changing times. Overall, the chapter supports the argument made throughout this book that British nature writing, from White to Woolf, is an intrinsically modern form that actively engages with the forces of modernity even as it often seeks to contest them. Against Bruno Latour's much-cited, if also much-misunderstood, 1993 pronouncement that 'we have never been modern', nature writing in Britain and elsewhere suggests that we have been modern, though not necessarily to our own advantage, for far longer than many of us like to suppose.[33]

The Birds and the Bees

(1) Seeing: Nicholson and Elton

For bird lovers, the twentieth century opened on a note of alarm. Between the closing decades of the nineteenth century and the end of the First World War, five birds of prey – the goshawk, the honey buzzard, the marsh harrier, the osprey and the white-tailed eagle – were declared extinct as British breeding species.[34] While habitat loss had played a significant part in their decline, the final decisive factor in their disappearance as breeding birds was direct human pressure: the persecution of birds of prey as vermin by farmers and gamekeepers, and the collection of birds and eggs as specimens. The speed with which one species loss followed another, along with the fact that a key factor in these losses was human persecution, led to calls for change from turn-of-the-century writers such as Hudson and Nicholson. Neither Hudson nor Nicholson were professionally trained ornithologists, but both self-consciously aligned themselves with emerging nature organisations: Hudson with the Royal Society for the Protection of Birds (RSPB), which was founded in 1889 and granted a royal charter in 1904, and Nicholson with the Nature Conservancy, which was established in 1949. Meanwhile, both men's work demonstrated the extent

[33] Bruno Latour, *We Have Never Been Modern*, trans. C. Porter (Cambridge, MA: Harvard University Press, 1993). Many an argument about modernity has been based on a strategic misreading of Latour, whose key contention is certainly not that modernity never happened, but rather that the so-called great divide between 'nature' and 'culture' made modernity possible while simultaneously falsifying it; there was thus a sense in which the idea of the modern depended (and still to some extent depends) on believing in the impossible necessity of an independent nature that exists 'out there'. The dismantling of this fallacy is central to contemporary new materialist approaches that see nature and culture as inextricably (and increasingly) entangled: an approach that this book largely follows, if also one it sees as in need of adjusting to different sets of social and cultural circumstances at different times. (See also the discussion of Latour in Chapter 1.)

[34] Over the course of the twentieth century, these raptors would reappear as breeding species in Britain, either through natural repopulation or human reintroduction. However, populations remain small and raptor persecution remains a significant problem.

to which nature writing, crossing over from the Victorian to the Modern period, informed the development of the scientific conservation movement and its institutions across the twentieth century and beyond.

A key feature of their arguments regarding bird life was their emphasis on the damage done even by those who purported to be appreciative of birds, whether aesthetically or scientifically. Nicholson followed Hudson in condemning 'the immeasurable losses which are being inflicted on the avifauna by [human] selfishness and greed – greed of eggs, greed of pheasants, greed of plumes, and greed of stuffed mummies'.[35] Nicholson reserved particular scorn for the recreational collection of bird specimens and eggs by private collectors. Specimen collection had long constituted an element of the scientific study of nature, but science was too often used as a pretext for amassing specimens that were little more than trophies. As Nicholson observed, 'It is the fatal logic of the collecting system that the more badly a bird needs carefully protecting, the more vigorously it is persecuted. The aggregate of birds' skins and eggs in collections is certainly not large, but the smaller the number of a species the greater the demand for specimens.'[36] Like Hudson before him, Nicholson recognised that the increasing systematisation of collection had merely increased its destructive capacity. Yet while some modern technological developments had increased the vulnerability of birds to man, others encouraged alternative approaches to bird life. In the 1907 edition of *British Birds' Nests*, for example, early wildlife photographers and brothers Richard and Cherry Kearton asserted the 'benefit which has accrued to the wild birds of our country by the introduction of this harmless and yet truly sporting method of studying Nature',[37] supporting their claim with a gushing testimonial from a reader: 'I consider that the birds ought to be extremely grateful to you for inventing bird photography. I never knew anything that has done so much for their protection during the nesting season as your example. I myself have for several years given up egg collecting entirely, and know many others who have done the same.'[38]

The desire to capture and preserve had not disappeared, but it was gradually being channelled into less destructive approaches to nature. Early twentieth-century animal protectionists argued for the need for

[35] Edward Max Nicholson, *Birds in England: An Account of the State of Our Bird-Life, and a Criticism of Bird Protection* (London: Chapman and Hall, 1926), pp. 189–90.

[36] *Ibid.*, p. 247.

[37] Richard Kearton and Cherry Kearton, *British Birds' Nests: How, When, and Where to Find and Identify Them* (London: Cassell and Company, 1907), p. vi.

[38] *Ibid.*, p. vii.

sanity and restraint in human engagements with nature. Nicholson presents Hudson as a representative of an emerging rationality in the approach to nature, praising him for the 'sanity of his outlook on Nature', his concern with the 'full and perfect presentation of the procedure of nature' rather than a vision of it as either 'a happy family' or 'a stupendous engine of torture'.[39] This characterisation can be seen, in part, as marking an attempt to differentiate twentieth-century bird (and other animal) protection from its Victorian antecedents. Nicholson was typically wary of sentimentality and anthropomorphism in the bird protection movement, dismissing the misguided 'super-humanitarianism' of those who argued that birds of prey should be shot as the 'Bird Foes of Birds'.[40] Seeking to uncouple the value of birds from human conceptions of morality, Nicholson argued for their integral place and function within a suprahuman natural order. Such conceptual and material factors enabled shifts in the approach to bird life during the early decades of the twentieth century, with J. C. Squire seeing fit to declare in 1928 that: 'Today there are a host of observers who watch birds with enthusiastic affection, never kill a bird, and would never dream of killing a rare bird. The modern man who kills a rare bird is not regarded as the hero of an exploit, but as a perpetrator of an unpunishable crime. The collectors, the hoarders of eggs, the stuffers of skins are now a furtive race.'[41]

Early twentieth-century protectionists argued that while modernity posed new threats to wildlife, modern attitudes might yet result in gains for persecuted species. Alongside concerns about intensifying pressures, both direct and indirect, upon bird and other animal populations and habitats, hope was expressed that an approach to nature informed by scientific understanding and guided by the principles of rational restraint could bring about a positive change in human relations with the natural world. One emerging field that gained purchase through the changing cultures of nature in the period was *ecology*. Arguing that a subject 'may change in character, not discarding the old, but rehabilitating it in a new synthesis', J. Arthur Thomson recommended a 'new natural history' informed by 'the great fact or idea of the inter-relations of organisms' and 'more or less equivalent to what is called nowadays – ecology'.[42] Early

[39] Nicholson, *Birds in England*, pp. 195–6. [40] *Ibid.*, p. 199.
[41] John Collings Squire, 'Introduction', in Ernest William Hendy, *The Lure of Bird-Watching* (London: Cape, 1928), p. 4.
[42] J. Arthur Thomson, *The New Natural History, Being the Twenty-Fifth Robert Boyle Lecture Delivered before the Junior Science Club of the University of Oxford on 6th June 1923* (Oxford: Oxford University Press, 1923), pp. 5, 4.

ecologists stressed both continuity and change in the view of the natural world their discipline promoted. In *Animal Ecology* (1927), a seminal work in the field of British ecology that synthesised developing concepts such as food chains, ecological niches and the pyramid of numbers, Charles Elton pronounced ecology to be 'a new name for a very old subject', selecting as his epigraph Gilbert White's views on the proper methods of the naturalist, whose work, for 'all that may be done at home in a man's study', is essentially to be done in the field, and ideally close to home.[43] While paying tribute to its long prehistory, Elton presented ecology as a discipline still in its infancy, observing several years later (1933) that ecologists had yet to trace the complete web of relations for even a single species, and suggesting that ecology in the early twentieth century now stood 'where geology [had been before it] at the end of the eighteenth century', on the cusp of becoming a science of wide cultural relevance but not quite having achieved that official status yet.[44]

One of the practical ways in which developing ecological knowledge affected the cultures of nature in early twentieth-century Britain was through the expansion of the concern for individual species into a concern for habitats with all of their characteristic features and component organisms. For example, the entomologist Charles Rothschild claimed in 1916 that it was 'useless to protect anything from the direct attacks of mankind when the home where it lives, or where it can live, is destroyed or partially destroyed by other causes'.[45] Rothschild argued for a better understanding of 'plant and animal association',[46] both for the sake of basic comprehension and in order to facilitate protection, stressing the interrelationship not only of natural organisms, but also of the disciplines dedicated to their study: 'It is impossible to separate entirely entomology from its sister science botany. If there were no plants, there would be no lepidoptera.'[47] Rothschild advocated for the establishment of nature reserves to protect characteristic or unique sites against the growing pressure upon their plant and animal populations. The announcement in *The Times* of the establishment of the Society for the Promotion of Nature Reserves in 1912 offered this admonition:

> That the larger wild creatures are steadily disappearing from the face of the earth with the advance of civilisation needs little demonstration. But

43 Charles Elton, *Animal Ecology* (Chicago: University of Chicago Press, 2001), p. 1. 44 *Ibid.*, p. 8.
45 Charles Rothschild, 'The President's Address', *Transactions of the Entomological Society of London*, 63 (June 1916), p. cxxxiv.
46 *Ibid.*, p. cxxxviii. 47 *Ibid.*, p. cxxxix.

> coincidentally with the wholesale extermination of mammals there is in process a no less disastrous process of destruction among the lesser creatures – birds, fishes, reptiles, insects, and plants.[48]

New naturalists like Nicholson, Elton and Rothschild also voiced a developing awareness of the distinction between natural and anthropogenic change. On one hand, Rothschild observed, 'The world is always changing; it always has been changing, and it always will change'; on the other, he drew attention to specific 'changes which are due to the actions of mankind' and argued that human intervention was necessary to arrest or limit the most egregious of these human depredations.[49] Rothschild's arguments made visible the tensions that characterised the nature reserve movement in Britain even in its earliest stages. He advocated for the identification and protection of sites that were still in a 'natural wild condition', but also considered 'quite artificial' the circumstances under which these isolated patches of land continued to exist.[50] In his argument that, in order to preserve the 'natural wild', some degree of control was necessary, Rothschild exemplified a discursive tension between the desire for idealised wilderness and the need for practical management that can be seen to traverse the entire history of the conservation movement in Britain, surviving into present times.

Elton features here again as a key figure. In 1957, thirty years after the publication of *Animal Ecology*, Elton delivered a series of radio talks on ecology that ranged from a consideration of the impact of invasive species to an argument for the conservation of species diversity, broadcasts later expanded into another full-length – and equally seminal – book, *The Ecology of Invasions by Animals and Plants* (1958).[51] Elton opened the first of his talks on an ominous note, declaring that 'Nowadays we live in a very explosive world', and – availing himself of the anxieties of the moment – likening the danger posed by explosions in the populations of plants and animals to that of nuclear blasts.[52] Elton presents ecological explosions ('the enormous increase in numbers of some kind of living organisms ... an infectious virus like influenza, or a bacterium like bubonic plague, or a fungus like the potato disease, a green plant like the prickly pear, or an animal like the grey squirrel')[53] as at once an inevitable by-product of modern society and a potential threat to human life. As he declares,

[48] 'Nature Reserves', *The Times*, 18 December 1912, p. 9.

[49] Rothschild, 'The President's Address', p. cxxxiii. [50] *Ibid.*, p. cxxxvii.

[51] Charles Elton, *The Ecology of Invasions by Animals and Plants* (Chicago: University of Chicago Press, 2000).

[52] *Ibid.*, p. 15. [53] *Ibid.*, p. 15.

'A hundred years of faster and bigger transport has kept up and intensified [the] bombardment of every country by foreign species, brought accidentally or on purpose ... from places that used to be isolated',[54] and has resulted in 'terrific dislocations of nature'.[55]

Elton does not present the distribution of species at a particular past moment as a fixed ideal. Noting that in previous geological ages, when the continents of the present day were still united in a single landmass, species were more widely distributed and 'cosmopolitan' than they are now, he observes that past natural events such as ice ages resulted in significant floral and faunal redistribution.[56] However, he emphasises the distinct character of contemporary anthropogenic change, arguing that in the modern era of global mobility 'we are artificially stepping up the whole business, and feeling the manifold consequences' of it, in both a positive and a negative sense.[57]

Anticipating recent arguments about species imbalances within the context of the Anthropocene, but without rehearsing some of their narrower nativist implications, Elton presents the redistribution of species as an unavoidable feature of the modern world.[58] Thus, while he concedes that 'we have to accept the proposition that invasions of animals and plants and their parasites – and *our* parasites – will continue', and that 'the balance between species is going to keep changing', he also maintains that environments can be made resilient in the face of new introductions.[59] He notes that, in addition to scenarios in which an introduced species establishes itself in a new environment, increases enormously, then disturbs, displaces or eliminates existing native species, a number of other scenarios exist in which introduced species fail to establish themselves successfully, or establish themselves 'without producing any noticeable disturbances or making our own similar species extinct'.[60]

Elton's work is seen today as foundational for the modern field of invasion biology, which seeks to understand why one species establishes itself and another does not, and why one species has a disruptive effect while another is absorbed without serious difficulties.[61] One way to explain this variation is through the protective function of variety in nature. 'The balance of relatively simple communities of plants and animals,' Elton

[54] *Ibid.*, p. 29. [55] *Ibid.*, p. 18. [56] *Ibid.*, p. 31. [57] *Ibid.*, p. 31.
[58] For a more extensive discussion of the Anthropocene, see Chapter 4.
[59] Elton, *Ecology of Invasions*, pp. 111–12. [60] *Ibid.*, p. 123.
[61] For an up-to-date study of invasion biology, see Cang Hui and David M. Richardson, *Invasion Dynamics* (Oxford: Oxford University Press, 2017); see also Daniel Simberloff, *Invasive Species: What Everyone Needs to Know* (New York: Oxford University Press, 2013).

remarks, 'is more easily upset than that of richer ones; that is, more subject to destructive oscillations in populations, especially of animals, and more vulnerable to invasions'.[62] He offers a range of examples to support this assertion, noting the vulnerability of isolated island ecosystems to disruption by introduced species and the greater frequency with which invasions and outbreaks affect 'cultivated or planted crops – that is, habitats and communities and species very much simplified by man'.[63] In contrast to the prevailing view of cultivated land as 'improved', Elton assesses arable land on ecological terms, grouping it with 'waste dumps, railway tracks, walls, and so forth' as simplified habitats vulnerable to further disturbance. In view of the vulnerability of simplified ecosystems, he believes there is 'something very dangerous about handling cultivated land as we handle it now, and even more dangerous if we continue to go farther down the present road of "simplification for efficiency"'.[64] Consequent upon this idea of the protective effects of diversity, Elton argues for 'some wise principle of co-existence between man and nature, even if it has to be a modified kind of man and a modified kind of nature': very much a conservationist mantra for our times.[65]

As one possible example of how variety might be maintained even in the already much modified landscape of the British Isles, Elton recommends the conservation of hedgerows and the vegetation of roadside verges: 'These long winding strips of habitat by the road and lane and field margins are the last really big remaining nature reserve we have in Britain, except for the wild moors and lakes of our northern mountains and the seas around us.'[66] He praises the richness, variety and stability of the communities that occupy these verges, and even welcomes the occasional disruption to these fringe systems, arguing that 'where road repairs and so on make a temporary disturbance', this 'only adds to the variety of ecological succession that can be seen'.[67]

In arguing for the conservation of hedgerows as one possible means towards the broader conservation of species variety, Elton departs significantly from the concern of early promoters of nature reserves with 'primitive wildness', to some extent anticipating the more recent work of the so-called new nature writers, who are as likely to celebrate 'semi-natural' places – that is, those that have clearly been modified by human interference – as those that remain in something like a state of the 'natural wild'.[68] Elton himself, however, is not usually categorised as a nature

[62] Elton, *Ecology of Invasions*, p. 145. [63] *Ibid.*, p. 145. [64] *Ibid.*, p. 145. [65] *Ibid.*, p. 155.
[66] *Ibid.*, p. 156. [67] *Ibid.*, p. 156.
[68] For detailed commentary on the new nature writing, see Chapter 4; see also Jos Smith, *The New Nature Writing* (London: Bloomsbury, 2017).

writer, despite his influence on other nature writers in Britain and elsewhere, notably the celebrated Americans Rachel Carson and Aldo Leopold. This seems in large part to be a matter of style. As Daniel Simberloff observes, Elton's characteristic style was 'dry and laconic, with occasional wry, sometimes sarcastic, humour but never emotional prose',[69] a fact which set his writing apart from more renowned nature writers of the time, who tended to display 'moving, almost poetic, rhetorical devices'.[70] This was evidently a conscious choice on his part. In a book review published in the *Journal of Animal Ecology* in 1939, Elton somewhat snidely remarks that 'we do not need metaphors about "fairy spears" to visualize the flashing mackerel in its course',[71] but this is not to be taken as a wholesale dismissal of nature writing, rather a declaration of affiliation to a particular, observational tradition of nature writing which, like several other naturalists of the period, he associates with Gilbert White: 'Will not the clear vigorous language of Gilbert White suffice to display the natural dignity and unsentiment [*sic*] of wild life in their cosmos?'[72]

Elton's homage to White is mainly on methodological grounds, and he is less concerned with emulating White's prodigious gift for accumulating descriptive detail than with matching his equally admirable ability to systematise his day-to-day, proto-ecological discoveries about human-animal (and animal-animal) interrelations in the natural world. As he recognises, though, White's prose is anything but dry, and while Simberloff's characterisation of Elton's own 'laconic' style is largely accurate, there is a liveliness about it that suggests that he is keenly aware of the need to attract as wide a readership as possible to his work. Indeed, the work of the new naturalists in general is more writerly than is frequently supposed, drawing on the discursive tradition of late nineteenth-century popularisers like Hudson and Richard Jefferies rather than replacing it, as Richard Mabey too hastily suggests, with the 'specialised and often know-all tones' of twentieth-century academic prose.[73] It might be going too far to describe the writing of the new naturalists as aesthetically self-aware, nor is it always particularly interesting to read, but it reveals some characteristics of what the contemporary literary critic Benjamin Morgan calls 'the aesthetics of data', in which quantitative description is accepted as an integral part of aesthetic experience rather than imagined to exist outside it in the shape of, for example, the modern statistical survey or the

[69] Simberloff, *Invasive Species*, p. 195. [70] *Ibid.*, p. 195. [71] Elton, *Ecology of Invasions*, p. 390.
[72] *Ibid.*, p. 390. [73] Mabey, *Oxford Book of Nature Writing*, p. 154.

professional scientific report.[74] Nor is such writing somehow metaphor-free, and Elton, possibly in spite of himself, displays a particular gift for figurative language that implicitly recognises its value in all forms of scientific communication, while he also more openly acknowledges that modern-day professional ecologists such as himself inevitably draw on everyday language, just as they rely on the 'observations of men like fishermen, gamekeepers, local naturalists, and, in fact, all manner of people who are not professional scientists at all'.[75]

If Elton's work is perhaps best characterised as popular science rather than nature writing per se, it also reminds us that what most of us take to be nature writing today emerged out of the popular natural-historical writing of the Romantic and Victorian periods, and that it continues to play a key popularising role in communicating usable forms of scientific knowledge about the natural world to a wide variety of publics, distributed on the continuum between lay and specialist. Above all, Elton, like other naturalists of his times, stressed the practical value of his work, which also fulfilled an important advocacy function in spreading a modern conservationist ethic among the emerging citizen-scientists of the British Isles.[76] By any account he was successful. In 1949, an act was passed providing for the establishment of national parks in the United Kingdom and a nature conservancy responsible for managing nature reserves and Sites of Special Scientific Interest (SSSIs). The institutionalisation of conservation on such a large scale was a substantial achievement, though threats kept pace with developments, and the post-war period saw the rise of new, indirect pressures on avian populations and other forms of domestic wildlife. For the second time in the century, British birds of prey appeared to be in danger of extinction, this time as a result of the introduction of synthetic pesticides that accumulated as they passed up the food chain, resulting in breeding failure and death in top predators such as the peregrine. The threat posed to birds of prey by these organo-chlorine pesticides was more insidious than that posed by direct persecution earlier in the century. While habitat remained, it was rendered increasingly toxic, destructive rather than supportive of life. In an age of toxicity and contamination, the interrelationship of organism and environment took on troubling implications, which were famously addressed in the United States by Rachel Carson and, in the United Kingdom, in J. A. Baker's

[74] For a contextualised discussion of the 'aesthetics of data', see Benjamin Morgan, 'After the Arctic Sublime', *New Literary History*, 47 (2016), 1–26.

[75] Elton, *Ecology of Invasions*, p. 6.

[76] Matless, *Landscape and Englishness*, p. 343.

less high-profile but, in some ways, equally startling work. The environmental movement was born, and conservation quickly became its media watchword, much of its protectionist discourse filtered through larger-than -life figures such as Jane Goodall and Gerald Durrell, both of them unmistakably English, but whose lives and works were conducted in an international limelight of their own making, just as conservation itself became increasingly unthinkable without the celebrity faces and voices, and the global media machinery behind them, that lent it institutional support.[77] This trajectory is picked up in the third and last part of this section; first, though, we need to account for two earlier figures, both of them celebrities in their own right, whose engagements with nature were predominantly a function of their *writing*: D. H. Lawrence and Virginia Woolf.

(2) *Sensing: Woolf and Lawrence*

Nature, argues the contemporary literary critic Bonnie Kime Scott, had a 'persistent presence' in British cultural modernism, although a metropolitan bias remained and some modernists, such as Ezra Pound, were keen 'to edit nature out'.[78] As previously noted, substantial scholarly work on 'green modernism' over the past decade or so has led to revisionist reappraisals of the role of nature in key modernist writers' work. Some writers looked for what Jeffrey Mathes McCarthy calls 'organic resolution ... through nature', while others claimed to have discovered a new appreciation for the agency of nonhuman nature and the interconnectedness of living creatures within the overarching (ecological) context of the web of life.[79] Included in both camps were writers interested in exploring the connections between human and *animal* subjectivities, contributing to a modernist tradition of animal writing that both connects to the wider genre of nature writing and constitutes a subgenre in its own right. Much of this writing, as Caroline Hovanec has shown, was scientifically informed, drawing on post-Darwinian understandings of animal emotion and vitality even as it continued to recognise that animals were in many ways fundamentally different from humans, and that their individual subjective experiences could never fully be understood.[80]

[77] The most accessible study of the relationship between conservation and celebrity is Daniel Brockington, *Celebrity and the Environment: Fame, Wealth and Power in Conservation* (London: Zed Books, 2009). See also the discussion of Durrell and Goodall later in this chapter.

[78] Scott, *In the Hollow of the Wave*, p. 16.

[79] McCarthy, *Green Modernism*, p. 2.

[80] Hovanec, *Animal Subjects*, p. 3.

An interesting pairing in this context is that of Virginia Woolf and D. H. Lawrence. Both can be seen, at least to some degree, as animal writers, but their views on animality differed sharply, with Lawrence's work being primarily concerned with the generative potential of animal drives and instincts, and Woolf's with the ways in which animal sense experiences participated in the process of making the world – multiple human and animal worlds – strange, both to each other and themselves. Despite (or perhaps because of) his experiences as a student of botany and a teacher of nature study, Lawrence's methods for exploring the natural world seem in direct contrast to those of the new naturalists.[81] Nicholson's recommendation of analytical distance and rational restraint in the patient accumulation of ecological knowledge is about as far as one can get from Lawrence's expressionistically rendered interspecies encounters and fantasies, which deploy a highly wrought, intensely personalised emotive language to immerse their readers in a colourful world of the senses shared, if not necessarily equally, by human beings and other living creatures alike. Lawrence's 1919 essay 'The Poetry of the Present', ostensibly an apologetics for free verse, is typical (insofar as anything is typical in Lawrence's work) for its attempt to capture the living 'now' while rescuing it from those narrow-minded biologists who, in their obsession to 'fix it with formation', succeed only in retrieving 'a hardened [fragment] of the past'.[82] These organicist views, which elevate holistic over compartmentalised knowledge, are also made clear in one of Lawrence's best-known essays, 'Pan in America' (1924). In the essay, Lawrence rejects the idolatrous modern worship of the machine, choosing to surrender instead to the 'vivid relatedness' that primordially exists between 'man and the universe',[83] and enjoining us (his fellow humans) to tap into a chthonic power which, together with the rest of living creation, draws spiritual as well as material sustenance 'from the dark bowels of the earth'.[84]

Grand pronouncements of this kind support a still influential view of Lawrence as a Rousseau-inspired modernist primitivist[85] whose work,

[81] M. M. Mahood, 'D. H. Lawrence, Botanist', in *The Poet As Botanist* (Cambridge: Cambridge University Press, 2008), pp. 183–225; L. Spolton, 'D. H. Lawrence – Student and Teacher', *British Journal of Educational Studies*, 14 (1966), 18–35.

[82] D. H. Lawrence, *Life with a Capital L: Essays*, ed. Geoff Dyer (London: Penguin, 2019), p. 78. The particular essays drawn on here are, in order of publication date, 'The Whistling of Birds' (1917, pp. 72–6), 'The Poetry of the Present' (1919, pp. 77–82) and 'Pan in America' (1924, pp. 194–206).

[83] *Ibid.*, p. 201. [84] *Ibid.*, p. 197.

[85] See, for example, Marianna Torgovnick, *Gone Primitive: Savage Intellects, Modern Lives* (Chicago: University of Chicago Press, 1990); also Dolores LaChapelle, *D. H. Lawrence: Future Primitive* (Denton: University of North Texas Press, 1996), which makes the case for Lawrence as a major

gloriously inconsistent though it is, celebrates the life force that flows through all living things, irrespective of their social circumstances, and which subscribes to what might be called today a new materialist view of the natural world in which all living creatures possess some degree of agency, and the vast sweep of creation is both immanent – that is, immediate to the senses – and transcendent ('The quick of all the universe, of all creation', Lawrence declares in 'The Poetry of the Present', is 'the incarnate self'.)[86] There is something to be said for this view although, as ever with Lawrence, an equal and opposite view may apply; as several commentators have pointed out, his work is powerful precisely because it presents an explosive collision of incompatible ideas and forces: at once an embracing celebration of newness and a destructive desire, in making space for that newness, to sweep everything aside.

Also in part keeping with the new materialism is Lawrence's view of nature as being 'in us, as well as without us'.[87] This view is possibly best expressed in one of his early essays, 'The Whistling of Birds' (1917), which, shuttling in typical modernist style between realist and symbolic registers, drifts in and out of the consciousnesses of the various natural phenomena it describes. Written during the First World War, the essay begins by inviting an analogy between fallen soldiers and frost-smitten birds, whose 'ragged bloody cloaks', scattered on winter ground, thinly cover the already rotting flesh within, which is mercilessly 'eaten by invisible beasts of prey'.[88] As so often in Lawrence's work, the dead soon give way to the living: the seasons change, nature revives itself, and other birds, first 'whistling faintly in the blackthorn thickets'[89] then emerging full force 'in the rising . . . silvery sap of a new summer',[90] announce the new order, in which life 'bubbles through' and they become like 'little well-heads, little fountain-heads whence the spring trickles and bubbles forth'.[91] As Lawrence implies, these birds, whether living or dead, are both us and not us. We may be like them in some ways, but we cannot influence them any more than we can alter the natural course of the seasons. 'We may not choose the world. We have hardly any choice for ourselves.'[92]

Nor, in other than some darkly intuitive way, can we *know* them any more than we can claim to know ourselves. This epistemic rift between human and animal worlds is probably explored most fully in Lawrence's poetry which, even as it looks to champion the physical strength and

twentieth-century nature writer. For the (considerable) influence of Rousseau on modern British nature writing, see Chapter 1.

86 Lawrence, *Life*, p. 82. 87 *Ibid.*, p. 74. 88 *Ibid.*, p. 73. 89 *Ibid.*, p. 73. 90 *Ibid.*, p. 74.
91 *Ibid.*, p. 74. 92 *Ibid.*, p. 74.

experiential authenticity of animals, is beset by communicative anxieties about them: is it possible to identify with them, or are these identifications little more than empty projections? Can we speak to them and, if so, do they speak back? As one might expect from Lawrence perhaps, many of these anxieties revolve around the sexual act. As Hovanec suggests, several of the animal poems in *Birds, Beasts and Flowers* (1929), notably the much-anthologised 'Snake', are best seen as provisional workings through of anxieties about human sex and individuality, although Hovanec optimistically sees them as prefiguring later, more assured accounts of animal sexuality as a 'primal, authentic expression of "blood consciousness"' which corrects the human tendency, deplored by Lawrence, to over-intellectualise the sexual act.[93] 'Snake' recoils from the same sexual instincts it invites. The eponymous snake, inserting his bullet head 'into that dreadful hole / And . . . snake-easing his shoulders / . . . entering farther', provokes a mixed reaction of fascination and disgust in the speaker who, previously intent on observing him, now feels powerless, either to kill him (as he thinks he ought to) or to follow him (as he secretly wants).[94] The autonomy of the snake, alternately likened to a king and to a god, merely reconfirms the impotent humanity of the speaker, a pattern repeated with variations in other poems in the collection. For example, in 'He-Goat', the goat/devil's frustrated sexual potency, stunted by his circumscribed life of domesticity, cannot help but remind the speaker of his own sexual frustrations.[95]

In poems like these, the sexual act is either proscribed ('Snake') or reduced to libidinous desire, the unconsummated experience of physical orgasm ('He-Goat'). In both cases, the speaker experiences animal sex only as simulation ('Snake') or as the displaced outpourings of his own sexual inadequacies ('He-Goat'). Identification with animals is only possible at this level through fantasised projections that simultaneously remind him of his incapacity to act upon his own animality, which the restraining force of human civilisation disallows. *Birds, Beasts and Flowers*, seen this way, offers sex education of a limited kind, though its birds and bees are sexually healthy in ways that human beings can only aspire to, in ways that human beings, for the most part, are emphatically not. Far from celebrating a 'primordial union between the thinking subject and the material world',[96] Lawrence's work often shows the unbridgeable gap between them. This gap looms particularly large in his animal writing, in which

[93] Hovanec, *Animal Subjects*, p. 128.
[94] D. H. Lawrence, *Birds, Beasts and Flowers* (Jaffrey, NH: Black Sparrow Books, 2007), p. 97.
[95] *Ibid.*, pp. 117–20. [96] McCarthy, *Green Modernism*, p. 202.

animal others, their differences only magnified by attempts to identify with them, display qualities conspicuously missing from deeply conflicted and often emotionally impoverished human selves.

If Lawrence's work focuses on sexual *health*, or the lack of it, Woolf's focuses instead on sexual *difference* as a set of potentially liberating alternatives to imprisoning social norms. Woolf uses animals to explore some of these alternatives, in keeping with her arch-modern views on the irreducibility of identity and her sustained imaginative explorations of the queerness of the natural world. Woolf's interest in animals and in nature more generally is apparent across her large body of work in which, as Derek Ryan sees it, 'immanent and intimate entanglements of human and nonhuman agencies are brought [repeatedly] to the fore'.[97] This suggests that her work, like Lawrence's, is amenable to contemporary new materialist approaches in which the humanist doctrine that 'holds only humans to be true subjects with real histories' is challenged, and nature and culture are inextricably entwined.[98] However, Woolf's representations of animals also need to be seen in the context of her time, as inventive responses to contemporaneous developments in the early twentieth-century life sciences, during a period that would witness a significant shift from the Victorian taxonomic tradition of natural history to the 'new biology' and the observation-based disciplines of ethology and ecology that would come increasingly to its forefront as the century progressed.[99]

Critics such as Gillian Beer, Louise Westling and Christina Alt have convincingly demonstrated Woolf's interest in ongoing developments and debates within the life sciences. In fact, she was not only familiar with Darwin – as were most writers of the time – but also conversant with early ecological ideas summarised by science popularisers such as H. G. Wells and Julian Huxley as well as turn-of-the-century popular writing on animal behaviour by Hudson, Jean-Henri Fabre and others, which prefaced the new, increasingly professionalised field of ethology represented in, inter alia, Nikolaas Tinbergen and Konrad Lorenz's

[97] Derek Ryan, *Virginia Woolf and the Materiality of Theory: Sex, Animal, Life* (Edinburgh: Edinburgh University Press, 2013), p. 4. As previously mentioned, Lawrence's work too has been seen from a broadly new materialist perspective: see, for example, Amit Chaudhuri, *D. H. Lawrence and the Poetry of the Present* (Oxford: Oxford University Press, 2003), which shows the intermingling of natural and cultural elements in his poetry; also Jeff Wallace, *Lawrence, Science and the Posthuman* (London: Palgrave, 2014), which as its title suggests focuses on the other-than-human aspects of his work.

[98] *Ibid.*, p. 150.

[99] Christina Alt, *Virginia Woolf and the Study of Nature* (Cambridge: Cambridge University Press, 2010), p. 53.

work.[100] This is not to turn Woolf into a closet new naturalist, still less to suggest that her work qualifies unproblematically as 'nature writing' in the sense in which the term is usually understood today. However, it does point to the fact that her work opened up a conversation with science, and that this conversation came to inform her experimental literary enquiries, many of them organised around human-animal and other ecological relations, into alternative ways of seeing and recording nature in light of what were increasingly recognised as the multiple contingencies of knowledge in the modern world.[101]

Issues around sexuality and gender were central to this conversation. Woolf's gendered critique of the masculinist biases of traditional (Linnaean) natural-historical systems of classification is well known, and need not be rehearsed here; less well known perhaps is that she saw modern biology as one possible outlet for the emancipation of women, with explorations of this possibility interspersed across the body of her work.[102] The most conspicuous characteristic of Woolf's literary-cum-scientific explorations of the natural world, however, is its *queerness*, which is connected to her own sexuality but by no means reducible to it ('queer' less as a descriptive noun or qualifying adjective than as a transitive verb).

One of Woolf's best essays on the queerness of the natural world is 'The Death of the Moth', published posthumously in 1942, which has been read in biographical terms as a displaced coming to terms with Woolf's imminent suicide. However, as always in her essays, it would be unwise to assume a homology between Woolf and her narrator, still more so given the predilection for assuming fictional identities that can be found in both her life and her written work. It is not that the narrators of Woolf's essays are unreliable; it is more the case that they are complicit with a body of work that mischievously shuttles between a series of shifting identifications and perspectives. These include *animal* perspectives, which at once indicate the potential for shared experience and confirm the impossibility of pinning experience down to any single perspective or limiting it to any single identifying category, whether human or animal, fictional or non-fictional, demonstrably living or indisputably dead.

Lepidoptera (butterflies and moths) are an obvious choice for such wide-ranging philosophical explorations. Classical and biblical associations of

[100] *Ibid.* Gillian Beer, 'Virginia Woolf and Prehistory', *Virginia Woolf: The Common Ground* (Edinburgh: Edinburgh University Press, 1996), pp. 6–28. See also Louise Westling, 'Virginia Woolf and the Flesh of the World', *New Literary History*, 30 (1999), 855–75.

[101] Alt, *Virginia Woolf*, p. 191. [102] *Ibid.*, p. 111.

butterflies with the soul have led some of Woolf's critics to see her fascination fascination with lepidoptera as belonging to well-established interpretive traditions,[103] but it is also important to recognise that, like Lawrence's, Woolf's animals are living and breathing world-historical creatures as well as multifaceted symbolic figures, and that a feature of her work, reflecting 'her long acquaintance with the study of nature, [is] its faithfulness to biological fact'.[104] In this context, it seems hardly surprising that the opening sentence of 'The Death of the Moth' doubles as an apparent statement of fact and a wry sideswipe at taxonomic classification: 'Moths that fly by day are not properly to be called moths.'[105] The brief narrative that follows, which describes the eponymous moth's increasingly desperate attempts to cling to life, reads similarly double: as a factual account of the moth's demise, rendered in dispassionate terms, and as a sympathetic, often lyrical attempt to identify with the moth's death struggle, in which 'a thread of vital light', becoming visible within 'his frail and diminutive body', is likened, in almost Lawrentian idiom, to 'a fibre, very thin but pure, of the enormous energy of the world'.[106] Each of the two versions destabilises the other, complicating interpretive attempts to read the account 'symbolically', but also equal and opposite attempts to read it 'biologically', as the study of one particular creature's transformation from living subject to dead specimen: the witnessed transition from one ontological state (life) to another (death).

One term for this unresolved perspective might be 'queer', and indeed the scene is described in terms of a 'queer spectacle', in which 'the queer feeling of pity' the narrator has for the moth contends with the brute fact that 'one could not help watching him', even as he is inexorably drawn towards his death.[107] The overtly gendered language of the account (empowered female observer, disempowered male observed) is challenged by the nearly imperceptible shifts it effects between personal and impersonal pronouns, as well as by its deployment of queer empathy, Nicole Seymour's capacious term for the imaginative capacity to forge 'empathetic, ethical interrelations between the queer and the nonhuman', suggesting that queer values, which attach in non-species-specific ways to immediate rather than familial relations, might also be 'the most effective

[103] *Ibid.*, p. 5. [104] *Ibid.*, p. 6.

[105] Virginia Woolf, *The Collected Essays of Virginia Woolf* (London: Read Books), p. 39. 'The Death of the Moth' appears on pp. 39–42; subsequent page references are to the collection.

[106] *Ibid.*, p. 40. [107] *Ibid.*, pp. 39–40.

ecological values' in the long run.[108] At the same time, Woolf's essay makes clear that the species boundary has not been crossed, with (generalised) 'strangeness' rather than (sexualised) 'queerness' becoming the operative term to describe the narrator's attempt to forge an empathetic connection with the animal other: 'Just as life had been strange a few minutes before, now death was as strange.'[109]

Others among Woolf's animal-oriented essays also benefit from being seen from a queer perspective. A good example is 'The Plumage Bill' (1921), originally written in response to attributions of blame for the failure of a parliamentary bill aimed at putting a stop to the early twentieth-century trade in tropical plumage, and usually considered today as an early example of the feminist polemic that would eventually build into more substantial interventions such as *A Room of One's Own* (1929) and *Three Guineas* (1938). The essay, along with the intemperate exchange it produced with one of the prime movers behind the bill, H. W. Massingham (journalist father of the previously mentioned H. J.), is trickier than it looks, with some contemporary commentators praising it for upholding Woolf's feminist principles and others viewing it more circumspectly as a critical faux pas in which Woolf's rage at Massingham's casual sexism – jointly revealed in his association of plumage with female vanity and his accompanying appeal to female tender-heartedness – involuntarily deflects attention from the main issue at hand, the indiscriminate torture of tropical birds.[110]

Reginald Abbott's detailed reading of the original text goes some way towards accounting for the various rhetorical devices around which it

[108] Nicole Seymour, *Strange Natures: Futurity, Empathy, and the Queer Ecological Imagination* (Champaign: University of Illinois Press), pp. 23, 27. Queer approaches to Woolf's work are unsurprisingly thick on the ground; less easy to find, however, are queer approaches to *nature writing*, despite the fact that there is a clear counter tradition of queer nature writing on both sides of the Atlantic, encompassing the earlier work of Thoreau and Melville (in the United States) and the more recent work of Derek Jarman and Philip Hoare (in the United Kingdom). See Robert Azzarello, *Queer Environmentality: Ecology, Evolution, and Sexuality in American Literature* (New York: Routledge, 2012), which includes a useful discussion of the writing of Thoreau and Melville; also Graham Huggan and Pippa Marland's 2020 article on Hoare in *ISLE*; see Graham Huggan and Pippa Marland, 'Queer Blue Sea: Sexuality and the Aquatic Uncanny in Philip Hoare's Transatlantic Eco-narratives', *ISLE: Interdisciplinary Studies in Literature and Environment*, isaa161, https://doi-org.eres.qnl.qa/10.1093/isle/isaa161 (accessed 14 September 2021). Although queer ecology has been commented on extensively over the past decade in, for example, the wide-ranging work of Stacy Alaimo and Catriona Sandilands, the links between nature writing and queer ecology have yet to be fully explored.

[109] Woolf, *Collected Essays*, p. 42.

[110] Virginia Woolf, 'The Plumage Bill', in *The Essays of Virginia Woolf*, ed. Andrew McNeillie, 6 vols. (London: Hogarth Press, 1986–2011), III (1988), pp. 241–5.

pivots. As Abbott recognises, Woolf's non-fictional essays often rely on the same playful but serious strategies of identity slippage (gender blurring, pet names, aliases and pseudonyms, etc.) for which her novels are celebrated. Nonetheless, he cleaves to the conventional view that while 'The Plumage Bill' raises 'the "woman question" [alongside] what might be called the "bird question"', Woolf herself 'distinguishes the two concerns as separate, with the woman question the more important of the two'.[111] While this view seems at first glance to be backed by Woolf's subsequent claim that she was writing 'not as a bird, or even a champion of birds, [but] as a woman', Abbott is at risk of compounding Massingham's original error by taking Woolf at her literal word.[112] In truth, Woolf is performing *both and neither* of these roles, and the implied queering of anthropocentric ('writing as a bird') and cisgender ('writing as a woman') perspectives in the original essay, as well as in the various critical exchanges it initiated, eventually yields another example of the challenge to fixed identity categories that Woolf presents throughout her work.

This isn't to claim that Woolf was hedging her bets, either as a pioneering feminist or as a forward-looking animal protectionist; rather it is to suggest that her work could serve to confound the expectations on which these fast-developing movements, both of which would pick up further pace in the following decades, were based. However, while Woolf may have been an iconoclast in several respects, she was also well aware of her debts to the past, not least in her conservationist leanings, and like many other conservation-minded writers of the time she felt obliged to pay her respects to the Reverend Gilbert White. *The Natural History of Selborne*, which she had read as a child, was duly revisited in a late (1939) essay in which she praised White for his detailed observational style and his sympathetic approach to his animal subjects, while still taking him to task for his less than reflexive stance towards his fellow humans: his deference to the nobility, for instance, or his not always forgiving attitudes towards the poor.[113] While she was not immune herself from this type of class critique, Woolf was certainly ahead of her time, as was White himself, in championing an informed approach to the natural world that was attractively readable in form and appropriately scientific in content, and that came increasingly to be aimed, not just at seeing or sensing, but also at actively

[111] Reginald Abbott, 'Birds Don't Sing in Greek: Virginia Woolf and "The Plumage Bill"', in Carol J. Adams and Josephine Donovan (eds.), *Animals and Women: Feminist Theoretical Explorations* (Durham, NC: Duke University Press, 1995), pp. 263–89.
[112] Woolf, 'The Plumage Bill', p. 245.
[113] Virginia Woolf, 'White's Selborne', in *Essays*, VI (2011), pp. 189–94.

saving threatened spaces (including the English countryside) and species (including native wildlife). This combination would become de rigueur for subsequent generations of conservation-endorsing British nature writers, albeit without seriously denting the social privileges upon which much of the modern history of British conservation, and the popular natural-historical writing that supported it, might be said to rest. Gerald Durrell is a key figure here, both as a hugely popular author and as a leading conservation practitioner, and it is to Durrell's work, and the mixed legacy it provided, that the chapter turns next.

Saving: Durrell and Goodall

Gerald Durrell is widely acknowledged to be one of the most accomplished nature writers of the twentieth century, both within and beyond Britain, as well as 'one of the great conservation leaders of the modern world'.[114] Although Durrell's star has arguably waned since his death in 1996, many of his books are still in print and *My Family and Other Animals* in particular remains popular more than sixty years after it was first published (1956). Similarly, Durrell's wide-ranging conservation work continues to be influential. The Durrell Wildlife Preservation Trust, established after his death, is one of the world's most recognised wildlife charities, while Jersey Zoo, one of his most treasured conservation projects, remains operational more than sixty years after it was first set up. During his lifetime, Durrell – a larger-than-life figure in almost every sense – identified himself as in the vanguard of the international conservation movement, for which, according to his biographer Douglas Botting, he believed himself to be one of the 'standard bearers in a new moral crusade' to save wildlife from the extinction that threatened it worldwide.[115] He also became a global celebrity at a time when conservation – now widely seen as a major conduit for celebrity – emerged as the media-driven, capital-intensive global phenomenon we recognise today.[116]

Durrell was a contemporary of David Attenborough's, and the two men shared several characteristics: a lifelong love of wildlife, a good understanding of the ways in which this love might be transmitted to a wide audience and a charismatic presence, both in front of the camera (Durrell, much like Attenborough, was a tireless, globetrotting broadcaster) and in other aspects of public life. Perhaps the biggest connection between

[114] Douglas Botting, *Gerald Durrell: The Authorised Biography* (London: HarperCollins, 2000), p. xvii.
[115] *Ibid.*, p. 240.
[116] See Brockington, *Celebrity and the Environment*; also Graham Huggan, *Nature's Saviours: Celebrity Conservationists in the Television Age* (London: Routledge/Earthscan, 2013).

the two, however, was (and, in Attenborough's case, remains) their self-presentation as *amateur naturalists*. Durrell, again like Attenborough, had few professional zoological credentials and never claimed as much; instead, he identified himself as an amateur in the British natural-historical tradition, a self-taught animal lover committed to educating others about the need to protect endangered animals and convinced of the intrinsic value of wildlife.

Durrell's focus on extinction seems up to the minute today, and in some ways he was a quintessentially modern figure, well aware of the uses that might be made of his own celebrity and by no means averse to mixing the profit motive of corporate capitalism with the moral imperative to save threatened wildlife. However, in other respects (and once again the comparison with Attenborough is apt) he was something of a throwback, in terms not just of his indebtedness to the natural-historical tradition, but also his cultivation of what might uncharitably be described as a colonialist sensibility in his life and his conservation work. The connections between conservation and colonialism are well documented,[117] while there is evidence to suggest that modern-day conservation, however well intentioned, serves predominantly First World political and economic interests, or provides the rationale for top-down forms of social and environmental control.[118] Durrell's conservation work is not unsusceptible to this critique, and it seems significant that much of it was done in the twilight of empire,

[117] See, for example, Richard Grove, *Green Imperialism: Colonial Expansion, Tropical Island Edens and the Origins of Environmentalism, 1600–1860* (Cambridge: Cambridge University Press, 1995); also Richard Drayton, *Nature's Government: Science, British Imperialism and the Improvement of the World* (New Haven, CT: Yale University Press, 2000).

[118] For a provocative approach and some striking examples, see Rosaleen Duffy, *Nature Crime: How We're Getting Conservation Wrong* (Manchester: University of Manchester Press, 2010); see also Bill Adams and Martin Mulligan (eds.), *Decolonizing Nature: Strategies for Conservation in a Post-colonial Era* (London: Routledge, 2002). While most scholarly accounts of conservation on a *global* scale pay at least some attention to colonial legacies, works on *national* conservation more specifically seem strangely reluctant to engage with this. David Evans' otherwise comprehensive account of the history of nature conservation in Britain, for example, has very little to say about the influence of empire or imperialism, while Bill Adams' 1996 study *Future Nature*, albeit focused on progressive agenda-setting, seems content to leave the issue of (British) colonial conservation history, together with its conspicuous inequalities, to one side. As Adams makes clear, conservation, whether of landscapes or wildlife, is in everyone's best interests, and if conservation, as he believes, is primarily 'about setting the terms of the engagement between people and nature' (p. 209), then those terms can only be agreed upon if as many of us as possible are involved. It is hard to disagree with this, but the fact remains that, in Britain at least, conservation has historically tended to reflect specific class interests in a country where around a third of the land is currently owned by the aristocracy. For a critique of the class-bound effects of British conservation, see Mark Cocker, *Our Place: Can We Save Britain's Wildlife before It Is Too Late?* (London: Cape, 2018). Crucial questions of ownership and access – with obvious implications for nature writing – are taken up elsewhere in this book, notably in its introductory and concluding chapters.

while his ancestral connections (he was born in India) link him clearly to the heyday of the Raj.[119]

A never seriously shaken belief in the superiority of the British also emerges periodically in his writing, especially in early works such as *The Overloaded Ark* (1953) and *The Bafut Beagles* (1954), which recount his African expeditions as a young man in language that cannot help but strike contemporary readers as racist, even if it was also very much a product of its time.[120] To some extent, this attitude also appears in by far his best known work, *My Family and Other Animals*, which in keeping with most of his early books is travel writing masquerading as natural history rather than the other way round. Certainly, the text has more in common with British travel writing of the time by the likes of, say, Eric Newby than it does with the natural-historical writing of his contemporaries. And like much of Newby's work, *My Family* – ostensibly a retrospective of the Durrell family's enchanted years in Corfu (1935–9) – offers a slapstick account of the misadventures of the English abroad, featuring many of that now dated subgenre's classic ingredients: locals with funny habits and even funnier accents, throwaway remarks about the 'native temperament' and, perhaps above all, a seemingly automatic deference towards the British: as the narrator says early on, tongue only half in cheek, 'The fact that I was English was sufficient [to be trusted], for the islanders had a love and respect for the Englishman out of all proportion to his worth. They would trust an Englishman where they would not trust each other.'[121]

My Family is redeemed to some extent by its instinct for self-parody and, more particularly, by its wry and at times even melancholy awareness of its own escapism, with its uncomplicated childhood perspective licensing old-school natural-historical pursuits which, even back in the 1930s, were well past their sell-by date. Ten-year-old Gerald is described gleefully stealing eggs and, in one case, snatching baby magpies straight out of their nest, while the island's bountiful natural environment reveals itself without too much effort and seems not to object over much to being disturbed. The childhood perspective also allows for exaggeration, with every family outing seen as an expedition and every new natural-historical discovery opening up a different (animal) world. People are included among these discoveries, and are assessed using similar behavioural

[119] Botting, *Gerald Durrell*, p. xvii.

[120] Gerald Durrell, *The Overloaded Ark* (London: Faber and Faber, 1953); *The Bafut Beagles* (London: Rupert Hart-Davis, 1954). See also Botting, *Gerald Durrell*, pp. 123–4.

[121] Gerald Durrell, *The Corfu Trilogy* (London: Penguin, 2006), p. 47.

criteria; often they are explicitly likened to animals and vice versa, as when young Gerald examines the courtship rituals of his teenage sister Margo's numerous Greek suitors, or introduces us to three visiting British artists, one of whom 'looked and sounded like a cockney owl with a fringe', and another of whom 'looked like a well-boiled prawn with a mop of dark, curly hair'.[122]

Durrell's gift for caricature is joined here to what might be called a juvenile form of sociobiology, the study of the biological basis for social behaviour which, a couple of decades later, would come to be associated with the Harvard-based entomologist E. O. Wilson's work. Mostly, though, *My Family* contents itself with shifting between 'older', natural-historical ways of describing and acting upon the natural world and 'newer', ecological perspectives. A good example is the chapter 'The World in a Wall', which reveals a wealth of life hidden within the crumbling wall of the Durrells' sunken garden. In scientific terms, this might be described as an ecological niche, a mixed biotic community of plants and animals, but scientific terms are beyond the self-taught Gerald, for whom the wall and its inhabitants are instead a 'rich hunting ground'.[123] Still, the descriptions the narrator provides are accurate enough and, anthropomorphisms aside, the chapter gives a good sense of the autonomy of a miniature world – one to which an impressionable young boy is granted privileged access, but also one which operates independently of him, its hybrid blend of predators and prey difficult for him to distinguish as 'everything seemed to feed indiscriminately off everything else'.[124]

This is nature writing in classic observational mode, yoked to a childhood idyll in which the 'secret' island of Corfu, still largely shielded from the outside world in the 1930s, offers a seemingly inexhaustible source of astonishment through its rich profusion of native wildlife. Nature functions alternately as spectacle and (adventure) playground, and often both of these at once, with the antics of its human and animal inhabitants amplified to the level of comic opera as the text, scattered lyrical interludes apart, moves at pace from one manic set piece to the next.[125] Though it wouldn't be published until a couple of decades later, one suspects that Durrell would have heartily agreed with the tone and spirit of Joseph Meeker's pioneering 1974 study *The Comedy of Survival*, where comedy, seen by Meeker as a 'mode of action common to many animal species and deeply rooted in evolutionary history', registers 'a valuable contribution

[122] *Ibid.*, p. 121. [123] *Ibid.*, p. 129. [124] *Ibid.*, p. 130. [125] *Ibid.*, p. 7.

to survival and a habit that promotes [individual and social] health'.[126] Comedy, for Meeker, means 'muddling through [rather than] progress or perfection';[127] oriented as it is towards the evolutionary advantages of cooperation and adaptation, it tends to avoid the types of polarised conflicts that characterise a tragic view of life.[128]

This might help to account for the fact that *My Family*, in keeping with most of Durrell's written work, is anything but a conventional conservationist vehicle; in fact, as Botting points out, 'there is not a glimmer to be found [in any of Durrell's early writings] of [his] views on the proper role of zoos, on animal conservation in general, and on the captive breeding of endangered species in particular', all issues that were central to his life's work.[129] It might also help to explain why Durrell's writing, allied to this work, was paradoxically effective in raising awareness of conservation issues, drawing attention to conservation precisely by *not* attending to it, and by choosing to focus instead on 'feelgood' accounts of the wonder and beauty of the natural world. Not that Durrell was shy of advertising the tragedy of extinction in the public domain, as evidenced in his typically forthright views at the 1976 World Conference on Breeding Endangered Species in Captivity: 'Mankind is creating false extinctions caused by overpopulation. All species are being pushed to one side by the damage we are doing to their habitats. Mankind continues to act as if there were another world just around the corner which we can use when we have ruined this one. But there isn't.'[130] As Botting observes, the sheer scale of animal extinctions worldwide would cast a long shadow over Durrell's life, which had its own fair share of losses. His father died of a brain aneurism when Durrell was just three, and much of his early writing emerged out of *that* shadow. The epigraph to *My Family*, taken from Shakespeare's *As You Like It*, references a 'humorous sadness' that is never far beneath the surface of his work; indeed, one of the many ironies that surround Durrell's life and work is that, while no one could have been more committed than he was to the protection of endangered animals, he was never particularly good at protecting himself.[131]

Durrell's writing, like Durrell himself, is marked by a restless energy that is always on the point of tipping over into excess and that routinely mocks its own extravagances. If its charms are faded today, that's probably because the Durrell brand is so obviously designed to be old-fashioned.

[126] Joseph W. Meeker, *The Comedy of Survival: Literary Ecology and a Play Ethic* (Tucson: University of Arizona Press, 1997), pp. 9–11.
[127] *Ibid.*, p. 160. [128] *Ibid.*, p. 37. [129] Botting, *Gerald Durrell*, p. 206. [130] *Ibid.*, p. 480.
[131] Durrell, *Corfu Trilogy*, p. 4.

In *My Family*, the anachronistic qualities of Durrell's work, filtered through a childhood persona, are strategically magnified, allowing him to present a view of a real place that exists partly in memory, but also partly in the imagination: a looking-glass world. The unreality of the text (as in the Alice books) is produced out of an exaggerated sense of innocence, supporting a light-hearted study of cross-species family ties in which, ironically, nonhuman animals are taken away from their families, seconded to a young boy's fantasy mission to bring as much of the natural world indoors as he can.

Published some fifteen years later (1971), Jane Goodall's *In the Shadow of Man* could hardly be more different in its approach, but its extended account of Goodall's pioneering *in situ* study of chimpanzee social relations – which would itself extend over a period of ten years – shares Durrell's basic aim of extending the human family to include other animals, while its aura of innocent idealism would feed into a 'Goodall legend' that proved every bit as durable as Durrell's, turning this unlikeliest of duos into the foremost celebrity conservationists of their times.[132] In Goodall's case, the legend emerged out of a timely conjunction of mutually influencing factors, notably her subject (primatology) and her gender (female): a combination which, almost unheard of at the time (the 1960s), would have a dramatically galvanising effect on the international media reception of her work.[133] Meanwhile, it can also be attributed to her celebrity status as a twentieth-century white British 'environmental saviour' who, albeit in a very different mould to Durrell's, became iconically associated with the global protection of endangered wildlife.[134]

Now well into her eighties, Goodall remains as popular a figure as ever, as do the various 'rehabilitant narratives' which, featuring chimpanzees and other apes as surrogates for humans, look to offer relief for the turbulent times we live in by rescuing endangered animals in order that humanity might be rescued from itself.[135] Conciliatory scenarios of human-simian contact are integral to these narratives, as they are throughout *In the Shadow of Man*, with one particular scene having become iconic in its own right. In the scene, Goodall, reacquainted a decade after establishing

[132] Jane Goodall, *In the Shadow of Man* (London: Weidenfeld and Nicolson, 1988).

[133] For a concise account of Goodall's rise to fame, see Linda Marie Fedigan, 'The Paradox of Feminist Primatology: The Goddess's Discipline?', in Angela Creager, Elizabeth Lunbeck and Londa Schiebinger (eds.), *Feminism in Twentieth-Century Science, Technology, and Medicine* (Chicago: University of Chicago Press, 2001), pp. 47–72.

[134] Huggan, *Nature's Saviours*, p. 7.

[135] Donna Haraway, *Primate Visions: Gender, Race, and Nature in the World of Modern Science* (New York: Routledge, 1989), p. 156.

contact with one of her original primate subjects, David Greybeard, holds hands with him in an intimate 'drama of touch'[136] that not only peels away the years, but also reaches out across differences of all kinds (sex, race, species) in a semi-mythical setting in which 'both are rehabilitated to a natural wildness to which Man, their ultimate stand-in, is [serendipitously] reconciled and restored'.[137] Donna Haraway reads the scene as emblematic of Goodall's 'message of natural peace',[138] harnessing it to what she sees as one of primatology's most striking mythic narratives, in which the figure of Woman, mediating between animal 'nature' and human 'culture', is conscripted for the task of mending the broken link between both of them and the master figure of Original Man.[139]

One of the points of such mythic narratives, Haraway suggests, is to blur the uneven relations of power that made the post–Second World War study of primates possible in the first place. Chimpanzees, *pace* Haraway, exist in the shadow of man in more senses than one, and the scientific breakthroughs of the 1960s – the discovery that chimpanzees and other apes were more like us than we could ever have thought – provided a powerful if also powerfully mystifying antidote to an era marked by the major social upheavals that accompanied decolonisation and the civil rights movement, and that was stalked by the potentially world-ending spectre of nuclear war. It is certainly true that popular scientific works like *In the Shadow of Man* convey little sense of the world outside of their own immediate sphere of influence, and it is particularly telling that Goodall's text makes so little of the coming into being of Tanzania as a fully independent nation in 1961, only a year after her own field studies began. However, at another level the text is seemingly aware of its own ambivalent status as a late colonial artefact. When Stephen Jay Gould, in attempting to rescue Goodall's work from the criticism it received at the time, rightly celebrates it as 'one of the great achievements of twentieth-century scholarship', he slightly spoils the point by denying that it is 'a dying gasp from the old world of romantic exploration'; in fact, the text derives much of its narrative power from the fact that it is an accomplished piece of modern scientific writing and a late Romantic travelogue *at the same time*.[140] It is also a typically anxious piece of animal writing, with its anxieties evenly distributed between Goodall's keen awareness of her own vulnerability in the face of her animal subjects, who are immeasurably more powerful than

136 *Ibid.*, p. 149. 137 Huggan, *Nature's Saviours*, p. 108. 138 Haraway, *Primate Visions*, p. 127.
139 Huggan, *Nature's Saviours*, p. 108; see also Haraway, *Primate Visions*, p. 150.
140 Stephen Jay Gould, 'Introduction to the Revised Edition', in Goodall, *Shadow of Man*, pp. vii–x (p. vii).

she is, and her nagging uncertainty about the validity of her scientific research, which asks more questions than it can possibly answer and is hedged about with qualifications of all kinds.[141]

To some extent, the care Goodall takes with the text, notably her reluctance to over-claim, is of a piece with the respect she shows the animals, who are socially and individually differentiated in ways that only an prolonged period of intense scrutiny allows. (Goodall is adamant throughout that her field research should not be mistaken for casual observation.) However, their surface similarities aside, the emotional lives of animals are *not* equated with those of humans, and while Goodall acknowledges that the lasting relationships some of them develop with one another 'may be [seen as] shadowy forerunners of human love affairs, I cannot conceive of chimpanzees developing emotions, one for the other, comparable in any way to the tenderness, the protectiveness, tolerance, and spiritual exhilaration that are the hallmarks of human love in its truest and deepest sense'.[142] 'Chimpanzees', she goes on to suggest, 'usually show a lack of consideration for each other's feelings which in some ways may represent the deepest part of the gulf between them and us. For the male and female chimpanzee there can be no exquisite awareness of each other's body – let alone each other's mind.'[143]

Remarks like this imply that while Goodall granted her animal subjects life histories of their own, she also saw them as lacking the intellectual and emotional complexity of their human counterparts; they also indicate that, like today's animal biographers, she was fully aware of the pitfalls of projecting human thoughts and emotions onto nonhuman animals, of attempting to account for animals' individuality without resorting to 'reconstruct[ing] their feelings [or] read[ing] their minds'.[144] Indeed, Goodall's support for conservation in *In the Shadow of Man* appears to be based on her recognition of animals' *differences* from humans: differences only further illuminated when, as is demonstrably the case with the great apes, they share several similarities with our so-called primitive ancestors as well as with our contemporary selves. Modern animal writing tends to support this view, which is based less on the alterity of animals than on the fact that we continue to know so little about them while so readily claiming some sort of emotional connection with them. And as twentieth-century writers as different in style and approach as Nicholson

[141] Goodall, *Shadow of Man*, p. 57. [142] *Ibid.*, pp. 122, 126, 196. [143] *Ibid.*, p. 187.
[144] On the fast-growing field of animal biography, see André Krebber and Mieke Roscher, *Animal Biography: Re-framing Animal Lives* (London: Palgrave, 2018); also Susan McHugh, *Animal Stories: Narrating across Species Lines* (Minneapolis: University of Minnesota Press, 2011).

and Jane Goodall would come to acknowledge, this gap between sympathy and knowledge was also a reflection of our continuing uneasiness with ourselves.

Country Files

If the animal narratives of the Modern period are not always immediately associated with nature writing, rural writing lies at its very heart. This is evidenced in what is usually acknowledged to be the first collection of British nature writing, Henry Williamson's *An Anthology of Modern Nature Writing* (1936), which focuses largely on the evocation of rural landscapes.[145] Indeed, as we move through the rural and place-based writing of the twentieth century, there is a growing sense of generic self-identification, with authors locating themselves in a tradition of nature-themed writing, both through their allusions to writers such as White, Jefferies and Hudson – and, across the Atlantic, Thoreau and Emerson – and through their interventions into the discourse of nature itself.

Notwithstanding this sense of a genre beginning to coalesce, in the Modern era place-based writing, like the landscapes and ways of living it evoked, underwent significant transformation. The changes were particularly marked in attitudes towards farming, which in the early part of the century was almost synonymous with nature, but which, in the wake of the intensification of agriculture after the Second World War and the deployment of pesticides such as DDT, came to be regarded as one of the greatest threats to it. This section of the chapter traces a broad arc from the complex pastoral of Edward Thomas' prose writings in the first two decades of the twentieth century to the flowering of agrarian writing in the 1930s and 1940s, a discussion that focuses on John Stewart Collis' *The Worm Forgives the Plough* (1973). We then proceed to the more self-consciously modernist perspectives of literary works such as Nan Shepherd's *The Living Mountain* (written in the closing years of the Second World War, though not published until 1977) and Margiad Evans' *Autobiography* (1943), before concluding with the post-pastoral reflections of Richard Mabey's *The Unofficial Countryside* (1973), which some see as inaugurating the 'new nature writing' examined in more detail in Chapter 4.

[145] Henry Williamson (ed.), *An Anthology of Modern Nature Writing* (London: Thomas Nelson, 1948 [1936]).

All of these works, to a greater or lesser extent, are preoccupied with images of the rural. By the end of the nineteenth century, for the first time, more of the British population lived in towns and cities than in the countryside. Yet, as Raymond Williams argues, while rural Britain was increasingly subsidiary to these burgeoning urban areas, 'so many of its ideas of how to live well . . . persisted' that 'there [was] almost an inverse proportion . . . between the relative importance of the working rural economy and the cultural importance of rural ideas'.[146] These ideas have been explored in relation to rural fiction of the time, but by comparison non-fictional rural writing has generated little critical attention, and is more often discussed as an adjunct to the fiction than considered in its own right. William J. Keith's *The Rural Tradition: William Cobbett, Gilbert White, and Other Non-fiction Prose Writers of the English Countryside* (1975) is one of few critical works to address non-fictional prose, and as Keith himself states, 'rural writing is a curiously neglected topic', despite its conspicuous popularity, which has proved remarkably durable over time.[147]

There are good reasons why rural life became such a focal point during this particular period. As mentioned earlier in this chapter, McCarthy locates the early decades of the twentieth century as a time that witnessed an acute yearning to articulate some sense of stability around British – and, more often, English – national identity, as ongoing political instability on the domestic front, combined with the cataclysmic effects of the First World War, rendered the national consciousness particularly fragile.[148] When McCarthy describes the era as 'fascinated by rural *England*' he speaks advisedly, given the predominance in its rural literature of the southern English counties of Wiltshire, Dorset, Sussex, Kent and Norfolk.[149] Nevertheless, the writing of the time also evidenced complex intersections with other nations and regions within Britain and Ireland, as well as a growing cultural attraction to the 'peripheries' of the Atlantic Archipelago. This centrifugal impulse was driven in part by an urge to reflect on the final throes of traditional ways of life that were proving impossible to sustain in the face of encroaching modernity, and in part by a desire to explore alternative, non-urban existences, the latter resulting in

[146] Raymond Williams, *The Country and the City* (London: Vintage, 2016), p. 356.

[147] William J. Keith, *The Rural Tradition: William Cobbett, Gilbert White, and Other Non-fiction Prose Writers of the English Countryside* (Brighton: Harvester Press, 1975), p. ix.

[148] McCarthy, *Green Modernism*, pp. 156–7. [149] *Ibid.*, p. 117.

a spate of island georgics.[150] It was also a time in which the varied landscapes of the British Isles as well as its fauna were deemed both worthy and in need of preservation. The period saw the founding of the Campaign for the Preservation of Rural England (CPRE) (1926), a sister body in Scotland, the Association for the Preservation of Rural Scotland (1927), and in Wales, the Council for the Preservation of Rural Wales (1928).[151] Some of the key authors studied in this chapter reflect the genre's bio-regional diversity: Edward Thomas, for example, was of Welsh ancestry and John Stewart Collis was born in Dublin, while Nan Shepherd drew on her long and deep acquaintance with the rugged mountain landscapes of Scotland's Cairngorms.

This geographical complexity was matched by the content and stylistic qualities of the literature. While some rural writing was harnessed to extremist political views (the classic case being Williamson's fascism), it is far too simple to see it as automatically tainted by 'blood-and-soil' politics, or to assume that its tone was one of unadulterated nostalgia, or, indeed, to reduce it to a compensatory outlet for the trials and tribulations of the modern world. McCarthy notes, for example, the radical potential of the agrarian novel, suggesting that modernist georgic opened up a site for the critique of capitalism.[152] Still, the rural genre was also complicit in dominant economic systems; it was a literature 'created in the main for urban consumption', in which the divide between city and country was becoming increasingly entrenched.[153] As will be seen in what follows, rural writing of the Modern period was marked by an equally sharp divide between leisure and labour, between the 'true' countryman/woman and the urban/suburban 'visitor', and by questions of class and education. At the same time, writers self-consciously engaged in reassessing existing aesthetics of the rural and in renegotiating their relationship with earlier literary works.

Edward Thomas' 'Gateway Vision'

London-born Edward Thomas was not a native countryman, but as a boy he spent time with relatives in Wales and in Richard Jefferies' county of

150 See, for example, R. M. Lockley's *Dream Island* (1930) and *Letters from Skokholm* (1947); also Frank Fraser Darling's *Island Years* (1940) and *Island Farm* (1943). For discussion of modernist 'island utopias', see Brannigan, *Archipelagic Modernism*, chapter 1.

151 The question of public access to these landscapes also came to the fore at this time, with the founding of the National Council of Ramblers' Federations in 1931, and the Kinder Scout mass trespass of 1932, which brought to a head frustrations over limited public roaming rights.

152 McCarthy, *Green Modernism*, p. 144.

153 Keith, *Rural Tradition*, p. 194.

Wiltshire. Jefferies, in fact, was a major literary influence on Thomas, who turned again and again to him for inspiration and solace.[154] His own biography of Jefferies, published in 1909, was dedicated to Hudson, whose work he also admired greatly. While Thomas' own writing career was cut short – he was killed in 1917, aged thirty-nine, on the opening day of the First World War Battle of Arras – he published a substantial body of rural-themed non-fiction in addition to biographies, reviews, fiction and poetry. His rural prose writing, which includes *Beautiful Wales* (1905), *The South Country* (1909), *The Heart of England* (1909), *The Icknield Way* (1913) and *In Pursuit of Spring* (1914), has not been as closely studied as his poetry, and critical responses to his evocations of the rural have not always been positive. Critics have tended to see him as relying on oversimplified versions of the classical pastoral, with its well-worn tropes of retreat (in Thomas' case from suburbs that, in his view, were 'so easy to make . . . sordid') and return.[155] It is true that such imagery abounds in Thomas' work, which references such idealised figures as the 'immemorial' peasant, and whose gestures towards a golden age are often accompanied by a routinised dismay at the passage of time. It is also true that some of his early works, in particular, lack a sense of locale. *The Heart of England*, for example, is presented in the form of condensed vignettes of place-based experience, with titles such as 'The Village', 'The Brook' and 'One Green Field', that speak to the emblematic and the iconic but, when it comes to actual geographical detail, remain strangely vague. Assessing *The South Country* in 1918, Herbert G. Wright similarly comments that it offers a 'composite picture' of landscape: 'It does not contain name after name in local order, nor does [Thomas] attempt to distinguish between one county and another.'[156] For Wright, this lack of differentiation contributes to Thomas' ability to conjure, as Thomas himself describes it, 'an ideal country, belonging to itself and beyond the power of the world to destroy'.[157] Keith is less convinced by this, seeing Thomas' idealisations as part of an 'artificial' and 'even pretentious' deployment of the pastoral, while Williams views his use of the pastoral as a wilfully adopted impediment, 'an eyeglass . . . lifted, deliberately and proudly, to the honestly observing eye'.[158]

[154] Edward Thomas, *The Childhood of Edward Thomas* (London: Faber, 2011 [1938]), pp. 134–5.

[155] Edward Thomas, *In Pursuit of Spring* (Toller Fratrum: Little Toller Books), p. 47; see Terry Gifford, *Pastoral* (Abingdon: Routledge, 2019), pp. 2, 47, 83, for a discussion of the retreat-and-return trajectory of pastoral.

[156] Wright, cited in Keith, *Rural Tradition*, p. 202.

[157] Edward Thomas, *The Heart of England* (London: J. M. Dent, 1909), p. 57.

[158] *Ibid.*, p. 57; Keith, *Rural Tradition*, p. 202; Williams, *The Country and the City*, p. 255.

However, what Williams interprets as deliberate and proud often seems, in Thomas' work, self-aware and implicitly critical. Many of Thomas' pastoral invocations are followed by qualifications. The itinerant flower seller and artist his narrator encounters on his way out of London in *The Heart of England* is gently satirised for his adherence to an idealised notion of the golden age, both in terms of his proclamation that there are 'no figures' in his paintings, and in his felt need for frequent trips to 'The Old Angel' or 'The Chequers' to refresh his vision.[159] Thomas' own evocations of the golden age in *The South Country* are likewise tempered by the observation that history and archaeology have debunked the idea of a past age of ease and felicity, and by reflections on the misleading power of the nostalgic impulse.[160] Each generation, he suggests, projects itself into 'backward' dreams of a more idyllic past, longing not for a particular historical epoch or historical region, but a 'region out of space and out of time'.[161] 'It is an odd world', he says, 'where everything is fleeting yet the soul desires permanence'; it is this oddness that his nature writing seeks to capture, sometimes in stylistically jarring prose.[162] As Robert Macfarlane remarks of *The South Country*, it 'slides without warning from the intensely observed to the extravagantly imagined', creating for the reader 'an intriguing cognitive dissonance brought about by [its] irregular movement between the empirical and the mystical'.[163] Or as Williams puts it, in Thomas' work 'intense observation of people and objects dissolves, without transition, into forms of fancy'.[164]

Whether viewed as failings, as they are by Williams, or as distinctions, as they are by Macfarlane, tension and apparent contradiction are the defining characteristics of Thomas' representation of the natural world. At times, Thomas suggests, the appeal of nature lies in its resemblance to the human world, as when he declares, 'The blackbird sings, composing phrases all the sweeter for being strangely like some in the songs that countrymen used to sing.'[165] However, a moment later he praises the *in*human quality of the nightingale's call, using it to dismiss the 'parochialism of humanity': 'Beautiful as the notes are for their quality and order, it is their inhumanity that gives them their utmost fascination, the mysterious sense which they bear to us that earth is something more than a human

159 Thomas, *Heart of England*, p. 14.
160 Edward Thomas, *The South Country* (Stanbridge: Little Toller, 2009 [1909]), p. 118.
161 *Ibid.*, p. 139. 162 *Ibid.*, p. 112.
163 Robert Macfarlane, 'Introduction', in Thomas, *The South Country*, p. 9.
164 Williams, *The Country and the City*, p. 255. 165 Thomas, *The South Country*, p. 42.

estate.'[166] Sometimes this switching of perspective is enacted within the space of a single phrase, gesturing towards a deliberate rather than accidental dissonance. In an extended paean to trees, for example, Thomas articulates his appreciation of 'their kindliness and their serene remoteness and inhumanity'.[167] A similar disconnect obtains between the indeterminacy of place in the early works and the wealth of botanical detail that accompanies its evocation, while elsewhere Thomas is able to move abruptly between idealistic and realistic registers in his descriptions of the people he encounters in rural settings. He depicts an itinerant umbrella repair man as conducting a 'pure wild existence', enmeshed in the natural scene through his physical resemblance to it – his crooked body 'like a branch of oak which comes straight with all its twistings', and his head of 'bristly grey hair like lichen' – while at the same time presenting him as eking out a living on the margins of society: an itinerant, precarious life.[168]

As in much writing of the Modern period, the material and the symbolic are closely intertwined here. Thomas' descriptions of rural folk as both shaped by and resembling their natural environments are inevitably reductive, gesturing by implication at their otherness from 'civilised' human society. However, they might also be regarded as an early expression of shared human/nonhuman materiality, an aspect of Thomas' work that, as Ralph Pite notes, anticipates aspects of the new materialism (see Chapters 1 and 4).[169] Ultimately, it is not a single view but a kaleidoscopic shifting of perspectives that is characteristic of Thomas' work – which in turn complicates the too-easy idea of him working in a borrowed pastoral register. Moreover, he is actively engaged in charting the tensions within his own work, for example, those generated by his exploration of the imaginative dimensions of his accounts of people and place and his simultaneous awareness that these are culturally mediated. Opening *The South Country* by acknowledging that the title of his book is borrowed from a poem by Hilaire Belloc, he signals the role literary antecedents play in shaping his sense of place, a role confirmed in later works such as *In Pursuit of Spring*, which features a succession of literary encounters with characters such as George Herbert, Philip Sidney and Thomas Hardy. Increasingly, though, Thomas expresses his impatience with his reliance on imported rhetoric, describing his education as having left him 'afflicted with ... English literature', or remarking of *The Icknield Way* that while he had originally

[166] *Ibid.*, p. 44. [167] *Ibid.*, p. 150. [168] *Ibid.*, pp. 164–71.
[169] Ralph Pite, 'Edward Thomas and Robert Frost: To Earthward', in J. Carruthers *et al.* (eds.), *Anticipatory Materialisms in Literature and Philosophy, 1790–1930* (London: Palgrave, 2019), p. 221.

set out to write a 'country book', suitably attentive to its surroundings, he has eventually only succeeded in writing 'another of those books made out of books founded on other books'.[170]

Thomas perhaps goes too far in such self-critique, especially since he reveals a complex understanding of the way in which landscape is materially co-created by culture and nature. In *The Icknield Way*, he notes that the earliest tracks humans followed were formed by animals, whose routes were in turn responsive to the conditions of the land.[171] Old roads of human design are subsumed under the same category of the natural, as 'the natural lines of travel for a travelling race'.[172] At the same time, Thomas presents these roads as imaginative and narrative creations, describing the Icknield Way as a product of antiquarian speculation ('History, Myth, Tradition, Conjecture, Invention') as much as physical foot traffic.[173] This mass of speculation works to unsettle understanding: 'still nobody knows or believes that anyone else knows' either the origin or the ultimate destination of the Icknield Way.[174]

Thomas' frustration at finding himself in the sway of 'other books' is counterbalanced by his evident belief not only in the inevitability of cultural mediation, but, more positively, in the power of imaginative literature to illuminate the natural world. His prose works often include passages from other writers as a means of giving a fuller sense of the landscapes in question; as such, he adopts a hybrid form in much the same way as earlier writers such as Charlotte Smith (see Chapter 1). His veneration for Hudson in particular demonstrates his conviction that nature writing at its best can play a vital role in interpreting the world, and even, *in extremis*, in preserving its memory:

> Mr Hudson so writes of birds that if ever, in spite of his practical work, his warnings and his indignant scorn, they should cease to exist, and should leave us to ourselves on a benighted planet, we should have to learn from him what birds were.[175]

Thomas' scenario is largely hypothetical, but it still strikes the twenty-first-century reader with a sense of horrified recognition; even as we read the nature writing of earlier centuries we are introduced to an abundance of nonhuman life – especially bird life – much of which can no longer be

[170] Edward Thomas, 'How I Began', in *Prose Writings: A Selected Edition, Volume 1*, ed. Guy Cuthbertson (Oxford: Oxford University Press, 2011), p. 178; Edward Thomas, *The Icknield Way* (London: Constable, 1913), p. iv.

[171] Thomas, *The Icknield Way*, p. 3. [172] *Ibid.*, p. 2. [173] *Ibid.*, p. 32. [174] *Ibid.*, p. 34.

[175] Thomas, *In Pursuit of Spring*, p. 187.

found in the United Kingdom. In this and other ways, Thomas' work foreshadows the melancholic strain of much contemporary nature writing, identifying it with the 'modern sad passion for nature' that was already apparent in his day.[176] 'It will soon have become impossible to see Nature for the books that are written about her', he exclaimed in exasperated response to what he saw as a positive deluge of nature writing in the early twentieth century, a sentiment now echoed a century later as our own sad passion for nature takes on new Anthropocene forms.[177]

Perhaps as a result of his critical engagement with books about the natural world, Thomas was cautious about his own deployment of the word 'Nature'. In a letter of 1908 to Walter de la Mare, he outlines his understanding of the term and corrects his own misuse of it:

> You ask me to define Nature. I used it vulgarly for all that is not man, perhaps because man contemplates it so, as outside himself, and has a sort of belief that Nature is only a house, furniture etc round about him. It is not my belief, and I don't oppose Nature to Man. Quite the contrary. Man seems to me a very little part of Nature and the part I enjoy least.[178]

Notwithstanding its note of misanthropy, this is an important clarification. Thomas recognises the danger of seeing human beings as separate from and superior to Nature, and demonstrates his understanding of Nature's constructed-ness. For Thomas, the word 'earth' is better and has a particular resonance. He sees the poet Meredith as an earth lover, an 'inhabitant of this earth', 'not so much an admirer and lover of Nature . . . as a part of her'.[179] Where the capitalised term 'Nature' implies distance between the human and the nonhuman, lower-case 'earth' is all-encompassing, suggesting an integral, mutual involvement.

These reflections are made in *In Pursuit of Spring*, which represents a significant development in Thomas' depiction of nature. As Keith notes, 'Gone is [Thomas'] attempt to transform the natural world into a verbal and symbolic construct which is only a dilution of the original.'[180] Keith singles out one particular example when Thomas, travelling across Salisbury Plain, describes how it 'makes us feel the age of the earth, the greatness of Time, Space and Nature; the littleness of man even in an

176 Edward Thomas, *The Country* (London: B. T. Batsford, 1913), p. 19, cited in Edna Longley, *A Language Not to Be Betrayed: Selected Prose of Edward Thomas* (Manchester: Carcanet, 1983), p. xvi.

177 *Ibid.*, p. xviii.

178 Edward Thomas, *Selected Letters*, ed. R. George Thomas (Oxford: Oxford University Press, 1995), p. 51.

179 Thomas, *In Pursuit of Spring*, pp. 57, 61.

180 Keith, *Rural Tradition*, p. 211.

aeroplane, the fact that the earth does not belong to man, but man to the earth'.[181] It is this insight that leads the ecocritic Laurence Coupe to present Thomas as a 'prophet of green studies', and Edna Longley to praise his prescient 'ecological vision'.[182] Longley argues that Thomas' reading of earlier writers including Gilbert White, George Henry Borrow, William Cobbett and George Bourne (Sturt) was part of 'an intensive effort to realize the point at which the rural prose tradition reaches *him*'.[183] Looking backwards was a starting point for taking the nature-writing tradition forward. This was no easy matter, however, for Thomas. As he wrote in a letter to Gordon Bottomley, 'There is no form that suits me, & I doubt if I can make a new form.'[184] Unlike White, Cobbett and even Jefferies, who moved from place to place in his life but maintained his sense of being rooted in his south country, Thomas recognised that he could never claim to be a 'countryman'. 'A man in the country must be a naturalist, an historian, an agriculturalist, or a philosopher, and I am none of them.'[185] As Robert Macfarlane argues, '[Thomas] is still often thought of as a pastoral poet, celebrating place and belonging, but . . . his true subjects are disconnection, discrepancy and unsettledness.'[186] Macfarlane is referring specifically to Thomas' poetry here, but these subjects can also be found in his prose.

In Pursuit of Spring shows evidence of Thomas beginning to dramatise his struggles with himself and with the problem of representation. His narrator figure, making his way from London to Somerset, is frequently interrupted by a strange character called 'the Other Man' who appears coincidentally at various points of his journey, often as an irritant to his own narrative. The character is partly based on Thomas' friend the Reverend Jesse Berridge, who accompanied him on some of the bicycle rides on which *In Pursuit of Spring* was based, but Alexandra Harris and others read him persuasively as Thomas' alter ego.[187] Given that Thomas' pilgrimage was not just to find the Spring but to pay homage to Coleridge

181 Thomas, *In Pursuit of Spring*, p. 116.

182 Laurence Coupe, laurencecoupe.co.uk/category/green-studies (accessed 24 November 2020); Edward Thomas, *The Annotated Collected Poems*, ed. Edna Longley (Hexham: Bloodaxe Books), p. 22.

183 Longley, *A Language*, p. xvi.

184 R. George Thomas, *Letters from Edward Thomas to Gordon Bottomley* (Oxford: Oxford University Press, 1968), p. 57.

185 *Ibid.*, p. 139.

186 Robert Macfarlane, *The Old Ways* (London: Hamish Hamilton, 2012), p. 25.

187 For discussion of Thomas' friendship with Jesse Berridge, including the identification of Berridge as the basis for the 'Other Man', see Anna Stenning, 'Edward Thomas's "Ecstasy": An Unpublished Essay', *Essays In Criticism*, 66 (2016), 466–87; for the 'Other Man' as Thomas' 'second self', see

in the place where he wrote 'Kubla Khan', the Other Man might also be regarded as a comic nod towards Coleridge's disruptive 'person from Porlock'. He clearly represents a kind of playfulness in the narrative, but he might also be read as a vehicle through which Thomas examines the difficulties of writing about nature. In one of their extended conversations, the narrator repeats the Other Man's digressions on the evils of notebooks, the keeping of which, he grumbles, distracts him from the 'great effects of Nature'.[188] His complaint coincides with the frustrations Thomas voices in the letter to Bottomley: 'O Lord, my dipping into 15 closely written notebooks for solid details to support my soarings and flutterings about the South Country'.[189]

In confessions like this, Thomas seems to suggest that writers cannot arrange things as they are in nature; they must aspire instead to an account of the spiritual elements of their encounters with the natural world. Thomas' admiration for Jefferies' *The Story of My Heart* gives a good sense of his belief in the power of nature to exert both a physical and a spiritual force, one that breaks down the barriers of linear time: 'Here for the first time was the whole man, brain, heart, and soul, the body and the senses, all that thought and dreamed and aspired in him. At every entrance the universe came pouring in, by all the old ways and by ways untrodden before.'[190] He also venerated Hudson's capacity, generated by 'the intense reality of his love' for the natural world, to pass 'out of this world into a mythic, visionary, or very ancient world'.[191]

An episode in *The South Country* expresses a similar transition. Walking along a rural road, Thomas sees through a gateway two young lovers 'gathering nuts in the warm sun, the silence, the solitude'. The scene conjures a trio of classical references (Helen, Demeter, Persephone), but also triggers the following reflection:

> For a day white-throated convolvulus hides all the nettles of life. Of all the delicate passing things I have seen and heard – the slow, languid, gracious closing and unclosing of a pewit's rounded wings as it chooses a clod to alight on; the sounds of poplar leaves striving with the sound of rain in a windy summer shower; the glow of elms where an autumn rainbow sets a foot amongst them; the first fire of September lighted among men and books and flowers – not one survives to compare with this gateway vision of

Alexandra Harris, 'Introduction to Edward Thomas', in *In Pursuit of Spring* (Toller Fratrum: Little Toller, 2016), p. 18.

188 Thomas, *In Pursuit of Spring*, p. 169. 189 Thomas, *Letters to GB*, p. 167.

190 Edward Thomas, *Richard Jefferies* (London: J. M. Dent, 1938 [1909]), p. 294.

191 Thomas, *In Pursuit of Spring*, p. 190.

> a moment on a road I shall never travel again. To rescue such scenes from time is one of the most blessed offices of books.[192]

The prose here may be mannered, but it also has depth and beauty. While the idea of a 'gateway vision' serves to reinforce Thomas' idea of himself as a wanderer, an outsider to the lived rural environment, it also introduces the idea of an access point to an intensity of experience that can never be repeated. The influence of the Romantic poets and the American Transcendentalists can clearly be felt here: Wordsworth's 'spots of time', for instance, or Emerson's 'transparent eyeball'. But Thomas is not fully drawn into the metaphysical, and while episodes such as this one may gesture towards transcendence, they ultimately remain within an earthly domain – the universe pouring in, rather than the human spirit pouring out – shot through with unresolved mourning for the 'delicate passing things' of the world. Just as Wordsworth's spots of time have been read both as refuges from and products of imperial expansion (see Chapter 1), it is possible to read Thomas' impulse to preserve the ephemeral and unrepeatable as informed by the concerns of his own era: high anxiety at the flight of time, combined with a deeply melancholic sense of the ultimate incomprehensibility of the world.

Thomas knows in his heart that a moment cannot be contained or taken out of time, yet he continues to be driven by the same yearning for permanence that infiltrates the awkward pastoral of the early works. A passage from *In Pursuit of Spring* shows him moving towards a lyric distillation of experience, simultaneously intimate and universal:

> Venus, spiky with beams, hung in the pale sky, and Orion stood up before me, above the blue woods of the horizon. All the thrushes of England sang at that hour, and against that background of myriads I heard two or three singing their frank, clear notes in a mad eagerness to have all done before dark; for already the blackbirds were chinking and shifting places along the hedgerows.[193]

It is a small step from these lines to Thomas' most celebrated evocation of a rural place and a moment caught in time forever, in the poem 'Adlestrop': 'And for that minute a blackbird sang / Close by, and round him, mistier, / Farther and farther, all the birds / Of Oxfordshire and Gloucestershire.'[194] Thomas' prose writings are usually seen as the precursor to the greater achievement of the poetry. But it is in the prose that the tensions of attempting to represent the natural world and his place in it are most

[192] Thomas, *The South Country*, p. 191. [193] *Ibid.*, p. 134. [194] Longley, *Annotated Poems*, p. 51.

clearly enacted. Precisely in 'not finding his form', Thomas perhaps *did* find a distinctly modern form that spoke to dissonance and dislocation, qualities he shared with other, more self-consciously forward-looking writers of his age.

Modernist Georgic: John Stewart Collis and the 'Gate of Labour'

The 1930s and 1940s saw an extraordinary flowering of agrarian literature, much of it now seen as classic British nature writing. This included A. G. Street's *Farmer's Glory* (1932), Claire Leighton's *The Farmer's Year: A Calendar of English Husbandry* (1933), Adrian Bell's *Men and the Fields* (1939), Fred Kitchen's *Brother to the Ox* (1939) and Henry Williamson's *The Story of a Norfolk Farm* (1941). As noted earlier, rural writing is often caught between the competing claims of leisure and labour, but what is significant about a number of these writers is that they themselves lived and laboured on the land. Irish-born John Stewart Collis, for example, spent almost six years as an agricultural labourer in Sussex and Dorset during and immediately after the Second World War, giving an account of those years in *While Following the Plough* (1946) and *Down to Earth* (1947), subsequently published together in 1973 as *The Worm Forgives the Plough* (henceforth *WFP*).

We concentrate on Collis' work here because it lies on the cusp of a significant alteration in the rural scene in the wake of the 1947 Agricultural Act. As Greg Garrard observes, 'Perhaps the most catastrophic changes in the British countryside [over the past century] have occurred in the ordinary agricultural landscapes of pasture, hay-meadow and arable field.'[195] While Collis writes self-consciously from within the 'Machine Age', his narrative predates the more rapid industrialisation of farming – and the realisation of its environmental impacts – in the decades that followed the Second World War.[196] In a retrospective note appended to the 1973 edition of *WFP*, Collis acknowledges that his accounts of rural life were valedictory in a way he did not fully anticipate at the time: 'The change of scene followed rather swiftly. Thus this book is about the last of its kind that can now be written in England' (p. 4). As such, Collis' agricultural writings function as a turning point for nature writing in the twentieth century.

[195] Greg Garrard, *Ecocriticism* (Abingdon: Routledge, 2004), pp. 43–4.

[196] John Stewart Collis, *The Worm Forgives the Plough* (London: Vintage, 2009 [1973]), p. 11. Further references are in the text.

In his preface to *WFP*, Collis explains his motivation for undertaking agricultural labour. 'I had hitherto regarded the world too much from the outside', he confesses, 'and I wished to become more involved in it' – a wish granted when he turned down an army post and turned his attention to the land (p. 3). The emphasis on labouring in *WFP* and its agrarian subject matter has led Robert Macfarlane to describe Collis as a 'writer of the land' rather than as 'a nature writer' (p. xii). While we would resist such a differentiation per se, it does help to illuminate the finer distinction between *pastoral* and *georgic* nature writing. As William J. Keith notes, 'the tradition of the georgic is manifestly closer to the spirit of the rural tradition than is that of the pastoral'; for all that, the georgic has received significantly less critical attention than the pastoral, partly because of its association with the domination of the natural world.[197] Virgil's *Georgics* brings the subjects of agricultural labour and war into conjunction, but also opposition: farming is a kind of armed struggle with nature, but it is also an art of peace and 'good living' that puts martial conflict into the shade. Collis situated his labour, as well as his writing, squarely within this tradition, noting when learning to harrow that 'while using this instrument I might just as well have been a contemporary of Virgil' (p. 27). He also recognised the scope that working on the land offered for different kinds of perception. 'Having opened the gate of labour', he writes, 'I had suddenly stepped inside the world, and could see objects with fresh eyes' (p. 20). This revitalised way of seeing offers a marked contrast to Thomas' 'gateway vision', which reflects an outsider view and a more indeterminate aesthetic. A later passage from *WFP* draws out a further distinction. Working late one day in the fields, Collis makes the following observation:

> The final part of the rick-making began when, the platform growing very narrow, I had to handle the sheaves with much circumspection. Down below I could see the roads becoming whiter and the fields darker and the woods more sombre, and as I glanced at them it occurred to me that perhaps after all this is how I would prefer to catch sight of beauty – through the corner of my eye, while immersed in something else, while not seeking it at all. (p. 138)

This calls to mind Thomas' distillation of dusk in *In Pursuit of Spring*, but offers a different aspect: a displaced georgic aesthetic in which beauty is glimpsed tangentially while the viewer is occupied with manual work.

[197] Keith, *Rural Tradition*, p. 5.

As with Thomas' pastoral, though, Collis' georgic is not without its contradictions. He was not born to the labouring class, but was from a wealthy Anglo-Irish family and educated at Rugby and Oxford. In his preface he freely accepts the label 'intellectual' in relation to himself, stating that he believes 'in the Mind more than anything else', and his narrative attests to his struggle to find acceptance among his fellow farm workers, though he is careful not to suggest that they themselves are without intellectual capacities (p. 3). Perhaps as a result of his dual status – as both labourer and self-confessed intellectual – Collis often finds himself in critical dialogue with existing literary discourses of the countryside, intervening to comic effect in the elevated discourse of Anglophone pastoral. In *WFP*, Wordsworth's preface to the *Lyrical Ballads* is gently satirised: 'Such are one or two of my reflections recollected in tranquillity after hoeing' (p. 20). The lofty opening of Thoreau's 'Walking', 'I wish to speak a word for Nature', is bathetically reframed in the first sentence of *Down to Earth* (the second of the two texts published as *WFP*) as 'I am anxious to say a word about the potato.'[198] Channelling Yeats, Collis confesses that after a day spent on a dunghill in an icy wind, 'I would long for [the] city and a clean warm room. I will arise and go now, I would say to myself most unpoetically, and build me a town, far, far away from Nature and all her winds' (p. 53). Light-hearted as these references may appear, Collis is keen to make the more serious point that the outsider view is a falsifying one. 'From the road a number of agricultural jobs look remarkably quiet, serene, slow, and easy', he explains, 'but if you stand beside the man in question you may find that he is putting out all his strength, is moving quite fast, and is in anything but a serene state of mind' (p. 26). Similarly, he challenges Hardy's propensity to imbue his rural characters with grand tragic narratives, arguing that the stories of 'the Judes of real life, the truly obscure Judes' end as they begin, in the fields (p. 33).

Collis is similarly circumspect in his assessment of the new kinds of farm machinery that were appearing at the time he was writing. His 1973 note on *WFP* makes elegiac mention of such fast-disappearing rural icons as haystacks and corn ricks, and he regularly brings out the satisfactions of skilled manual labour in his written work. But he certainly does not object in principle to modern technological developments, welcoming them where they alleviate the toil of agricultural labour, and only condemning them when they result in the destruction of farming communities and the

[198] Henry David Thoreau, 'Walking', available via www.gutenberg.org/ebooks/1022 (accessed 15 January 2021); Collis, *WFP*, p. 165.

loss of jobs (pp. 65, 146). He extends this nuanced consideration to the subjection of farm animals to forms of technology. The mechanisation of milking, he suggests, is an extension of the already unnatural activity of milking by hand. 'Cows', he notes, 'were never made originally to be milked by hand . . . or to be milked by man at all' (p. 64). He recognises the way in which the animals themselves have become natural-cultural hybrids, manipulated by selective breeding for human benefit. '[It is us who] have put those big bags on the cow' (p. 64). Similarly, he notes that sheep are 'now half created by nature and half by man', countering classical and biblical associations of the composite figure of shepherd-and-flock with the visceral, anti-pastoral insight that rather than 'searching for a dear lost sheep, and rejoicing over it when found [the shepherd] is more often discovered searching [it] for maggots, and finding plenty, whole rings of them worming their way through the wool, the skin, through the flesh to the very bones' (p. 62). At the same time, Collis realises that farm animals too stand to benefit from the advent of mechanisation. Working with a horse-drawn chain harrow, he comments: '[T]he horse gave me the impression of hating it – though I may be wrong. It struck me forcibly, and to my surprise, that here was the job of jobs for a machine, a tractor' (p. 32). For Collis, the tractor releases horses 'from their slavery', and he questions the motives of those who would argue against this, accusing them of the 'proverbial cruelty of your sentimentalist' who is concerned with 'the spectacle of horses in the field' rather than with the 'horse as a living creature in its own right' (p. 66). Again, this observation offers a useful corrective to the pastoralising view from the road, which underestimates the sentience and suffering of both human and nonhuman labourers.

As these observations suggest, Collis' apprehension of rural life and landscape is a reflexive one. He dismantles any assumed superiority of the countryman, who, he writes in a section suitably entitled 'Unromantic View of Agriculture', is often 'inhuman, callous, violent, and cruel' (p. 51). He also challenges the urban/rural divide by asserting that it is the countryman himself 'who dreamt the dream of conquering Nature and of escaping from her. [It was he who] built London and New York' (pp. 107–8). There is something rather wilful about these assertions, as if Collis is testing but does not really believe his own hypothesis. *WFP* is certainly not didactic in georgic terms – Collis is too much the ingénu to instruct us in ploughing and harrowing – but it does suggest a preferred kind of human existence. Prefiguring the opening of Carson's *Silent Spring*, Collis describes the malaise of the new city dwellers: 'But as time went on a strange unease began to afflict the people in the towns. As they walked through the everlasting streets they began to pine for the open field, for the blessed sun, for the realities and simple joys they had left behind' (pp. 108–9). Ultimately, for Collis, an agricultural existence that is also able to incorporate the life of the mind is the ideal (p. 95).

Collis' self-definition as an intellectual does not necessarily stretch, however, to a full embrace of science. In a by-now familiar gesture of early twentieth-century nature writing, he declares his appreciation for scientific accounts of natural phenomena even as he maintains his conviction that nature cannot be wholly encompassed by scientific explanation. The 'classification business', he says to illustrate his point, is an attempt to 'overcome the problems of reality by simply labelling [it]': an attempt that 'only makes the thing more mixed than ever' (pp. 185, 264, 185). Art is viewed as similarly incapable of comprehensive representation. The surface of a tree trunk remains

> something that cannot be told, cannot be rendered. Here is the object, the thing itself, so staggering in its presence that we fall back from it. The intricacy of the totality cannot be copied, and it is the intricacy that is the picture; before it the art of suggestion is powerless . . . we must leave our pen, our brush in the face of it, abandon art as a hopeless substitute. (p. 265)

The inevitable inadequacy of representation, however, does not distress Collis. Instead, a vein of wonder runs through his work, as it does for so many other early twentieth-century nature writers, and he declares himself content to 'stand cap in hand' before the phenomena of the natural world (p. 207). This contentment extends to a form of spirituality, albeit an earthbound one. Towards the end of *WFP*, Collis gives full rein to his sense of enchantment in a passage of intensely lyrical observation which conjures up an image of insects and foliage all glittering in the sunshine. It brings him to exclaim: 'suddenly I realize where I am! I am in the Garden of Eden' (p. 233). The epiphany gives rise to a reflection on the biblical creation myth in which he interprets our departure from Eden as 'the birth of consciousness' which rendered us '*onlookers* . . . separated from Nature' (p. 233). Nevertheless, in his view, a further extension of consciousness can return us to the Garden, 'For it still exists – all we need is the key of the gate' (p. 233). The conduit to such extended consciousness is, like the key to the gate itself, found in labour.

For Collis, such redemptive georgic endeavour is not restricted to the human domain. He openly appreciates photosynthesis as 'the labour [of green plants] upon which rests the life of the world and the destiny of nations', and, like White and Darwin before him, praises the 'remarkable works' of the earthworm, a 'monarch among animals' who performs 'vast operations in ploughing, in harrowing, in levelling, in draining, in airing, in manuring, and even in creating the soil' (pp. 197, 171). Equal attention is given to processes of composition and decomposition. Observing maggots 'composing themselves while decomposing [a dead] bird', he reflects, 'The bluebottle is necessary. The bluebottle is good. All things in nature have a meaning and a purpose' (pp. 247, 246). Statements like this hark back to the natural theology of an earlier era while also reflecting a post-Darwinian sense of the commonality of animal life (see Chapter 2). The humble maggot, for example, is a representative of all animals which, unlike plants that subsist by 'eating only the air, only the earth', require 'the death of another that they might live' (p. 246).

Similarly, the dissolution of the material world is to be welcomed. Recounting the end of a typical day working on the land, Collis writes, in an earthy epiphany that resonates with contemporary new materialism, of catching a 'sharp whiff of smell coming from the upturned earth', which he identifies for us as 'fertility ... the *odour* of eternal resurrections from the dead (p. 154). As with his location of Eden in the everyday present moment, this insight provides an alternative reading of Christian doctrine. Here is eternal life, he argues, here, in the fertile matter of the soil. This sense of earth-bound wonder is all the more apparent when Collis comes to consider the existence of fossils. Holding the flint cast of a sea urchin conjures in him an awareness of an ancient continental and archipelagic connection in which 'Australia and New Zealand, joining with Africa and South America, made a single mass. The north of England and Ireland were one with America and with Norway' (pp. 187–8). At the same time, Collis registers a deep-time consciousness that diminishes the importance of the human. Fossils are textual, 'books of stone' that enable him to grasp 'the actual existence of that monstrous age to which they joined [him]' – an age which includes the evolution and subsequent disappearance of the dinosaurs, and the hundred million years between their demise and the eventual appearance of Man (pp. 191–2).

In passages like these, Collis seems ahead of his time, anticipating the Anthropocene consciousness of the present. But was he also *of* his time? One of the more curious elements of the book, given that it records Collis' experiences on the land for almost six years beginning in 1940, is how little a part the Second World War plays in the narrative, even featuring in some throwaway comic asides. In one of these, he recounts a moment of pastoral longing as he views a distant field and observes, in mock tribute to Remarque, that 'all [is] quiet on the agricultural front' (p. 81). True to his georgic theme, Collis seems to be suggesting that the agricultural fields of England are where the real struggle is taking place. His rare explicit mentions of the war do register it unequivocally as a tragedy, symbolised by the death of farm worker Reggie, who is killed when he handles an unexploded bomb, naïvely believing it to be inactive: 'The catastrophe of our time was focused upon the body of this one man, cut down suddenly in the midst of abounding life' (p. 270). Nevertheless, the closing pages of *While Following the Plough* (the first of the two volumes combined in *WFP*) offer an extraordinary, distinctly Virgilian vision of the endurance of the land, and work on the land, in the face of human folly. Looking across the fields, Collis also looks 'across the centuries' (p. 157). He then waxes lyrical: 'This was the eternal tale. This did not alter and would not stop ... Those men below at work upon the thresher, whom I have not sought to glorify – yet are they not glorified in these natural tasks? – support the conditions for the theatre of history, but they work outside the drama' (p. 151). In this context, the military forces passing by are almost an irrelevance: '[T]hey passed in long lines, these the chariots of our day ... History [itself] simply passed us by' (p. 158). There is an uneasy sense of fatalism here as well as a misplaced confidence in the supporting role of agriculture in the bigger theatre of history. Ironically, the modest farming life Collis celebrated would come to be a key player in the still unfolding drama of the Anthropocene (see Chapter 4).

As the second half of the twentieth century wore on, a great schism opened between those who championed nature and those who worked the land. Who after Collis could confidently write: 'A farmer is a liberator of the energy in the earth, ceaselessly creating what is good, and adding on a vast scale to the beauty of the world' (p. 35)? Certainly by 1947 the great flowering of agrarian literature had had its day, and it would dwindle from this point onwards. Georgic endeavour was outstripped by the hollow victory of mega-farms, by the wholesale destruction of hedgerows Elton had warned of (see previous section), and by the use of toxic farm chemicals whose catastrophic impacts were recorded by the likes of J. A. Baker (see Chapter 4). But Collis' preoccupation with how to see nature and how to live well off the land ensured a lasting influence for *WFP*, which found more commercial success in the 1970s than it had ever done in the late 1940s. At the same time, the fascination with the materiality of nature and preoccupation with new ways of seeing human beings' place in the natural world that are the twin characteristics of Collis' writing were taken in new directions by two innovative female nature writers whose work spanned the middle years of the twentieth century and to whom we now turn.

Earth Writing: Nan Shepherd and Margiad Evans

The rural literary tradition of the twentieth century is predominantly male, with the genres of pastoral and georgic allying themselves with melancholic and virile forms of masculinity, respectively. Thomas' rural women, as we have seen, are quickly subsumed into classical tropes, while Collis is uncomfortably dismissive of female aspiration: 'For whatever unfeminine feminists may preach, women, the most sophisticated as well as the most simple, invariably say sooner or later – "All I want is to make a home"' (p. 112). We now turn to two female nature writers of the mid twentieth century, in whose work any notion of a restricted female sphere (a domestic realm) is immediately gainsaid by their conjuring of an immense *oikos* (a planetary home) that reaches into the earth and out into space. The two authors in question are the Scottish writer Nan Shepherd, whose *The Living Mountain* was written in the closing years of the Second World War, though not published until three decades later (1977), and Welsh Marches writer Margiad Evans, whose *Autobiography* first appeared in 1943. Both were novelists and poets as well as prose nature writers, who received acclaim for their work at the time of its publication, but whose reputations declined even in their own lifetimes such that their work subsequently fell out of circulation. Evans has yet to achieve full recognition, especially for her non-fictional prose, though interest in her work is growing both within and beyond her adopted Wales, while Shepherd is increasingly seen as one of the most important nature writers of modern times.[199]

If Collis' work, as discussed previously, can be seen in terms of georgic-inflected 'land writing', then Shepherd and Evans represent a kind of 'earth writing' in which 'the material presence of the earth [is integrated] with the activity of making it into meaningful text'.[200] Shepherd and Evans are notable for their desire to engage with the earth in all its material

[199] The first full-length ecocritical study of Shepherd's work, Samantha Walton's *The Living World: Nan Shepherd and Environmental Thought* (London: Bloomsbury, 2020), identifies Shepherd as an inherently ecological writer whose work is highly relevant in the contemporary context of environmental crisis. The Welsh presses Parthian, Honno and Seren are gradually reprinting Evans' works, including her novels, and she is the subject of the critical collection by Kirsti Bohata and Katie Gramich (eds.), *Rediscovering Margiad Evans: Marginality, Gender and Illness* (Cardiff: University of Wales Press, 2013).

[200] Philipp Erchinger, 'Introduction: Earth Writing', *Philological Quarterly* 97 (2018), 125–35 (p. 125).

dimensions, but also for their willingness to reflect on the difficulty of articulating it in textual form. Wrestling persistently with language, and adopting a wide range of frequently experimental perspectives, these writers' vision enacts its own earth-centred heuristic, but this is not to say that the writers themselves are not also deeply involved in specific locales. *The Living Mountain* derives from Shepherd's long familiarity with the Scottish Cairngorm massif, while Evans' *Autobiography* is set in a small area of the Welsh Marches near Ross-on-Wye. Nor are we suggesting that these writers' gender renders them somehow 'closer to the earth', or that they are writing in opposition to what has gone before them. The perspectives of the Romantic poets, the American Transcendentalists and, in the case of Evans, the Victorian nature writer Richard Jefferies all resonate through their work, as they do through that of Thomas and Collis.[201] Rather, what is distinctive about their writing is the heightened sense of an embodied relationship with the earth and of the encompassing planetary consciousness that derives from it. This shared awareness would in turn help Shepherd and Evans to establish 'a new posthumanist pattern' which, challenging 'androcentrism and the nature/culture binary' and fostering fresh forms of relational thinking, arguably emerged during the Modern period, even if it is now more commonly associated with the work of the ecofeminists and new materialists of our times.[202]

The power of Shepherd's writing derives not only from her deep knowledge of the Cairngorms, but also from the varied, often distinctly unusual perspectives from which she seeks to view her chosen terrain. Consider the following recommendation:

> Lay the head down, or better still, face away from what you look at and bend with straddled legs till you see your world upside down. How new it has become! From the close-by sprigs of heather to the most distant fold out of the land, each detail stands erect in its own validity. In no other way have I seen of my own unaided sight that the earth is round. As I watch, it arches its back, and each layer of landscape bristles – though *bristles* is a word of too much commotion for it. Details are no longer part of a grouping in a picture of which I am the focal point, the focal point is everywhere. Nothing has reference to me, the looker. This is how the earth must see itself.[203]

[201] Evans' admiration for Thoreau and Jefferies is noted by Karen Caesar, 'Warding Off the Real: The Recreation of Self in *Autobiography* and *A Ray of Darkness*', in Bohata and Gramich, *Rediscovering Margiad.*

[202] Scott, *In the Hollow of the Wave*, p. 2; see also Chapter 4.

[203] Nan Shepherd, *The Living Mountain* (Edinburgh: Cannongate, 2011), pp. 10–11.

This passage pushes at the conventions of nature writing in a number of ways, endorsing defamiliarising perspectives that playfully deflate the importance of the human observer. The gaze suggested here is not that of the self-privileging (male) mountaineer, striking masterful poses on the summit, but that of the strategically decentred (female) viewer, whose intimation of an earthly intelligence allows her both to imagine and, via a plural form of self-consciousness, to participate in the broader planetary processes by which the earth experiences itself. This embodied perception is integral to whatever we might provisionally term 'earth writing', which is characterised by its extraordinary openness to the nonhuman and by its concern, not just with articulating, but actively *becoming* the earth.

Margiad Evans shares Shepherd's attempt to capture – but also her awareness of the inherent difficulty of capturing – a sense of the earth experiencing itself. '*I would see as the light sees*', Evans writes in her curiously named *Autobiography*,[204]

> *I would know as the earth knows when it feels the grass pulled by the sheep, when it rounds the rabbit's nest, when the blow of the hoof shoots to its heart. I would feel as the water feels round the rocks and the roots and the breasts of the swans.*[205]

For Evans, this knowledge cannot come from formal study, but must spring from personal experience. 'I'm not a naturalist or a scholar in leaves and birds', she writes, 'but something is there which makes me stand quite still and look'.[206] Rather than standing still in order to see the world, Evans, like Shepherd, often finds new perspectives in alternative arrangements of her body, for example, lying on the ground to see the stars wheeling above her as well as the tiny flowers in the grass beside (compare John Clare, whose writing often represents flora and fauna viewed from ground level). Such explorations allow her to distinguish between 'thoughts in the brain' and 'thoughts in earth', giving greater weight to the latter when it comes to the creation of 'good, true and *loving* earth writing'.[207] Evans' formulation here is appropriately ambiguous: 'earth writing' can be read as a noun or, with only a slight shift, as a present

[204] Evans' *Autobiography* shares few of the characteristics one might normally associate with the autobiographical genre. As Steven Lovatt notes, although written from a first-person perspective, it covers only a few years of her life, and it is deliberately shorn of the solipsism that marks most autobiographical texts. See landlinesproject.wordpress.com/2020/05/11/how-margiad-evans-wrote-the-earth-by-steven-lovatt/ (accessed 4 February 2021).

[205] Margiad Evans, *Autobiography* (London: Arthur Barker, 1952 [1943]), p. 36. [206] *Ibid.*, p. 56.

[207] *Ibid.*, p. 62.

continuous form in which we see, at one and the same time, Evans' autobiography reaching beyond her and the earth writing itself.[208]

Relational thinking of this kind derives at least in part from the ecological nature of both Shepherd's and Evans' work. Evans demonstrates a deep-seated sense of interconnection and shared materiality when she writes, 'One lives plurally . . . One is in everything. One lives throughout the universe and beyond. Though I think what I see I don't see with my eyes alone . . . All existences merge, even physically. Things more than resemble one another, they are more than kin, they are one manifestation.'[209] Shepherd, for her part, offers a more scientifically informed account of the ecological relationship. Although it was only in the 1977 foreword to *The Living Mountain* that she would refer to ecology directly, the original text is explicitly concerned with the 'intricate interplay of soil, altitude, weather, and the living tissues of plant and insect' that are ecological components of the mountain environment she describes.[210] The book also proceeds broadly along the lines of an ecological survey, describing the abiotic features of the landscape before turning to its biotic characteristics. Halfway through, Shepherd pauses to remark:

> I have written of inanimate things, rock and water, frost and sun; and it might seem as though this were not a living world. But I have wanted to come to the living things through the forces that create them, for the mountain is one and indivisible, and rock, soil, water and air are no more integral to it than what grows from the soil and breathes the air. All are aspects of one entity, the living mountain. The disintegrating rock, the nurturing rain, the quickening sun, the seed, the root, the bird – all are one.[211]

Humans are not excluded from this ecological picture. Throughout *The Living Mountain*, Shepherd takes care to acknowledge the lived and worked dimensions of the landscapes she describes, landscapes which dispel any sense of the Cairngorms as an uninhabited wilderness. For all her initial impression that, standing on the plateau and casting around, 'man might be a thousand miles away', Shepherd makes extensive reference

[208] This resonates with another 'earth writer' of the Modern period, Jacquetta Hawkes, whose idiosyncratic, formally complex geological history *A Land* (1951) is now seen, alongside *The Living Mountain*, as a classic example of modernist nature writing. For a useful discussion of the text, which is seen as anticipating current Anthropocene thinking, see Hayden Lorimer, 'Memoirs for the Earth: Jacquetta Hawkes's Literary Experiments in Deep Time', *Cultural Geographies*, 19 (2012), 87–106.

[209] Evans, *Autobiography*, p. 124. [210] Shepherd, *Living Mountain*, p. 59. [211] *Ibid.*, p. 48.

to those who currently live and work on the mountain – from climbers to crofters, from scientists to gamekeepers – while she also stresses past human presence in the region, finding it in the dotted remains of shepherds' huts, the paths scored through a much-visited and worked-upon landscape, and ancient Gaelic place names.[212] Similar attention is paid to changing patterns of land use in the region and their accumulated effects, such as the destructive effects of deforestation, for example, the removal of vital habitat for local species, some of which are in seemingly terminal decline.[213]

Similarly, Evans' Welsh Marches are mixed sites of human community and endeavour marked by changing land use and patterns of post-pastoral obsolescence: 'Now the old people [are] dead and the young [are] orange-farming in South Africa. At the farm, cars and machinery and no more maids walking up the hill late on a summer's evening to fetch the best water from the spring.'[214] Evans' relationship to the land is at least in part a georgic one: she writes of growing vegetables, of gathering and chopping firewood, and she takes work in the fields from time to time to augment her income from writing. Like Shepherd, she has a strong sense of attachment to the land, which is built up by intense and repeated interaction with it, but also like Shepherd, she is aware of the limits of her own knowledge of it, whether this is scientifically designated or experientially derived.

Evans and Shepherd also share a conviction that there would be much more to learn about the natural world if only we had the modes of perception to do so. As Shepherd remarks, 'There must be many exciting properties of matter that we cannot know because we have no way to know them.'[215] Our perceptions, inevitably shaped as these are by sensory limits and cultural conventions, thus need fine-tuning, which in turn 'deepens [our] sense of outward reality'.[216] Such fine-tuning involves a greater understanding, not just of how perception works, but also of the different ways in which it can be distorted. For example, atmospheric conditions in the mountains affect visual perception, producing optical illusions. A peregrine in the mist seems fabulously magnified, while 'solid ground that seems three steps away ... lies in sober fact beyond a 2000 foot chasm'.[217] At other times, the three-dimensionality of objects appears exaggerated: viewing a distant croft in the rain, Shepherd states that 'Rain in the air has ... the odd power of letting one see things in the round, as though stereoscopically.'[218] Reminiscent of her vision of the curvature of the earth, this sense of seeing things 'in the round' enlarges

212 *Ibid.*, pp. 76–7. 213 *Ibid.*, p. 80. 214 Evans, *Autobiography*, p. 12.
215 Shepherd, *Living Mountain*, p. 105. 216 *Ibid.*, p. 11. 217 *Ibid.*, p. 42. 218 *Ibid.*, p. 42.

Shepherd's apprehension of landscape. While perceptual distortions of different kinds 'drive home the truth that our habitual vision of things is not necessarily right', their disorienting effects are ultimately energising, suggestive of a vibrant world acting upon her senses rather than lying passive beneath her gaze.[219] Evans is similarly fascinated by the tricks and quirks of perception, and by the way in which they emphasise her own implication in the scenes that she is seeking to describe. She sees the wintry air as 'charged with surging whirling particles', and asks, 'Are they seen or a part of sight? Is this light-dust eternal dancing in the deeps of the sky or in the deeps of my own eye?'[220] For Evans, as for Shepherd, the confusion is invigorating. 'I cannot tell', she confesses in response to her own question, 'but when I stare upwards into the blue zenith the sparkling air divides itself into unspeakably bright atoms ever moving'.[221] Whether or not this vision of dancing particles is an optical illusion, it conjures a powerful sense of a universal material animacy that is not usually available to the perceiving subject.

It should be clear by now that both writers feel a marked personal connection with the natural world, not in an egotistical or solipsistic way, but in a manner that extends their sense of involvement with the earth. As Samantha Walton notes in relation to Shepherd, understanding the affective impact of the natural world upon us helps us to 'appreciate that we are a part of it'.[222] Shepherd finds in the landscape a kind of companionship, experiencing the greatest sense of communion when she has 'gone out merely to be with the mountain as one visits a friend with no intention but to be with him', and writing in her 1977 foreword of such communion as 'a traffic of love'.[223] Evans too feels a sense of reciprocity with the earth that is marked not just by sympathy, but also by love: 'I know that there is in me a growing sympathy. Between the ground I walk on and the centre of my breast, there is love.'[224] The key to such feelings is a heightened receptivity to the material world. 'The senses must be used', Shepherd exclaims, stressing how for her, 'eye and touch have the greatest potency'.[225] 'Touch is the most intimate sense of all', she maintains in almost erotic mode, insofar as 'the whole sensitive skin is played upon, [and] the whole body, braced, resistant, poised, relaxed, answers to the thrust of forces incomparably stronger than itself'.[226] Although Shepherd stops short of ascribing sentience to the mountains themselves, Evans goes

219 *Ibid.*, p. 101. 220 Evans, *Autobiography*, p. 72. 221 *Ibid.*, p. 72.
222 Walton, *Living World*, p. 197. 223 Shepherd, *Living Mountain*, pp. 15, xliii.
224 Evans, *Autobiography*, p. 56. 225 Shepherd, *Living Mountain*, pp. 96, 98. 226 *Ibid.*, p. 102.

further, exploring ideas of sensuous reciprocity in her observation that 'When the rivers flow, when the twigs move over the sky, when the wind touches the skin of the water, you can see the sensations they bring to one another, and your own body reflects them.'[227]

While there is a carnal intensity to Shepherd's writing, there is also a spiritual dimension to her work. This is mirrored in Evans' writing, where self and world are unified in an intensified experience that borders on the mystical. Walton has written of Shepherd's interest in Theosophy and Buddhism, and Shepherd herself claims to understand 'why the Buddhist goes on pilgrimage to a mountain', concluding that 'The journey is itself part of the technique by which the god is sought. It is a journey into Being; for as I penetrate more deeply into the mountain's life, I penetrate also my own.'[228] Despite the mention of god, this is a chthonic, earthy form of spirituality in which divinity is immanent in the body of the mountain and in the corresponding response of the pilgrim. Similarly, walking in the dusk, Evans finds her own access to the mystical. In an episode that chimes with Thomas' distilled vision of an evanescent moment and Collis' beauty glimpsed through the gate of labour, she writes,

> As I went along I made an effort to *climb out* and get in these things [the dusk, birdsong, a broken twig] – into the mysterious darkening and sealing of the earth, the quietening that is as the loveliest psalm of rest. And at last I did. I stood leaning on a gate. I was behind the sky. I was in the ground. I was in the space between the trees. My meaning grew in the earth and the firmament – I in the Nothing in which all is related.[229]

The idea of climbing out in order to get in reflects the dynamic between self and world that is at the heart of much modern nature writing. The assertion of the 'I' for Evans, and of being and self for Shepherd, might be viewed as an anthropocentric resurgence of the human ego, or even as a move towards transcendence. But what is striking about Shepherd and Evans' mysticism, if it can be called that, is that their spiritual feelings are not separate from their earthly incarnations; they arise from a sense of total material and spiritual interrelationship.

This is not to say that these two writers ignore the harsher phenomenologies of such involvement. Shepherd displays an acute awareness of the power of the suprahuman forces that surround her on the mountain – those

227 *Ibid.*, p. 91; Evans, *Autobiography*, p. 36.
228 Walton, *Living World*, pp. 12–13; Shepherd, *Living Mountain*, p. 108.
229 Evans, *Autobiography*, p. 51, emphasis in the original.

ungovernable elements (wind, rain, snow) against which the fragile human body may prove powerless to act.[230] Her 1977 foreword attests to the strength of these forces, revealing a preoccupation with the horror of death in the mountains: a child is killed falling from a chairlift; a man and a girl are found 'months too late . . . the girl on abraded hands and knees as she clawed her way through drift'; a lone skier goes missing; a group of schoolchildren are found dead.[231] Evans' *Autobiography* also reveals darker moments that reveal a more personal sense of haunting and of division from the self. 'Who can be that benighted person walking on the bank so fast alone?' she asks of herself at one point. 'It was my shadow in the moonlight like an impression in wax of my own ghost.'[232] This spectral effect is accompanied by a debilitating loss of purpose, particularly in relation to her writing. She compares herself to 'a very old woman knuckling [her] hands in [her] lap and repeating forgotten deaths' as she tries to gather the threads of what she wanted to put down on paper.[233]

As this troubling image indicates, even the moments of the most profound sense of connection can be disorientating, and the translation of experience into language a major challenge – one that gives rise to modernist expressions of self-consciousness. Evans, for example, feels that her visions have words while she is walking,

> But when I come in – oh I can't explain. I want words. More words for colour – for the blue which is beyond the grey sky; for another blue which the hills keep for days and which is *substantial* in a disappearing earth – and another word for haze or mist which is not dimness any more than the stars not shining in their nonentity. I want words which open, words for space, words which *will not bend the thought*. Is there such language? I believe there is. I believe I hear it. I aspire.[234]

But just as with the ephemerality of contact with the natural world, such aspirations frequently end in frustration. In part, Evans suggests, this is because of the multistranded immediacy of experience, which belies the linear progression of written description: 'All is simultaneous [hence] the incommunicable wonder, the difficulty of writing . . . Consciousness has many layers, profundity on profundity. For this one and separate miracles, the universe, the now, substitute words which must be added one after

[230] Shepherd, *Living Mountain*, p. 80. [231] *Ibid.*, p. xlii. [232] Evans, *Autobiography*, p. 17.

[233] *Ibid.*, p. 17. Evans' feeling of being haunted and separated from herself has been written about in the context of her epilepsy: a result of the brain tumour that led to her untimely death at the age of forty-nine. See A. J. Larner, 'A "Herstory" of Epilepsy in a Creative Writer: The case of Margiad Evans', in Bohata and Gramich (eds.), *Rediscovering Margiad*, pp. 129–41.

[234] Evans, *Autobiography*, p. 25.

another, and the vision finishes as a list.'[235] Such loss of simultaneity is mirrored in the form of *Autobiography*, which moves between sporadic diary entries and more sustained passages. Evans also reflects, in typically modernist style, on the impossibility of communicating with other species. Her childhood belief was that 'there is a consciousness in all living things', but as the demands of the adult word encroach, she finds herself losing 'not the sense of it, but the hope that it might be intercepted and partly understood'.[236] As she acknowledges, with deep regret, 'Wherever I go I must carry with me speech in my mouth and brain which is as uncomprehensible [*sic*] to the other kinds of being as the sparrows' jangling is to me.'[237] Even in trying to communicate with her own (human) kind, Evans – much like Shepherd – finds words woefully inadequate. There is thus a dialectical sense in both writers' work whereby knowledge is gained by being with the earth, but in which the resulting insights are hard to maintain, even harder to translate and part of an always unfinished process. As Shepherd wisely tells us, 'The thing to be known grows with the knowing', but knowing itself is an endless task.[238]

Coda: Richard Mabey's Post-Pastoral

Towards the end of her aforementioned geological history *A Land* (1952), Jacquetta Hawkes reflects on one particularly sorry example of a post-industrial British landscape, arguing that abandoned areas of this kind have now assumed 'a degraded ugliness never before known'.[239] 'Who can ever express the desolation of these forlorn scenes?' she writes: 'The grey slag heap, the acres of land littered with rusted fragments of machinery, splintered glass, tin cans, sagging festoons of barbed wire; vile buildings, more vile in ruin; grimy stretches of cement floors, shapeless heaps of broken concrete.'[240] Such 'scenes' might be seen as an ultimate realisation of the fears that Thomas had already voiced at the beginning of the century, when he suggested that the suburbs could be made so easily sordid. Seen in this light, the last text featured in this chapter, Richard Mabey's *The Unofficial Countryside*, first published in 1973, presents a kind of reverse trajectory. Far from fleeing the suburbs in search of nature, as Thomas does, Mabey purposefully walks into them, intent on relishing the unexpected flourishing of flora and fauna he finds there. The countryside Mabey describes is not 'countryside' at all in any conventional sense of the

[235] *Ibid.*, p. 75. [236] *Ibid.*, p. 76. [237] *Ibid.*, p. 76. [238] Shepherd, *Living Mountain*, p. 108.
[239] Jacquetta Hawkes, *A Land* (London: Harper Collins, 2012 [1951]), p. 193. [240] *Ibid.*, p. 193.

word; it also offers a provocation to limiting definitions of place more broadly. The locations he writes about – flooded gravel pits, rubbish tips, reservoirs, canals, golf courses, churchyards, urban gardens and sewage works – 'have one quality in common, and that is that, in them, the labels "urban" and "rural" by which we normally find our bearings in a landscape, just do not apply'.[241] The narrative that ensues emphasises the ways in which nature and (human) culture intersect unexpectedly in such inherently uncategorisable places: 'I have called it the unofficial countryside', Mabey writes, 'because none of these places is in the countryside proper, nor were they ever intended to provide bed and board for wildlife. They are all habitats which have grown out of human need.'[242] At the same time, these inadvertently produced nonhuman habitats are salutary reminders that life goes on, offering a corrective to blinkered anthropocentrism and the priorities of capitalist modernity.

Mabey's critical approach to the existing rural tradition acts as a kind of hinge, looking back to the twentieth century and beyond, but also forward to the twenty-first century (see Chapter 4). 'Our [prevailing] attitude towards nature', he complains, 'is a strangely contradictory blend of romanticism and gloom. We imagine it to "belong" in those watercolour landscapes where most of us would also like to live. If we are looking for wildlife we turn automatically towards the official countryside, towards the great set-pieces of forest and moor.'[243] Such wild landscapes are not needed, he insists, in order to engage with nature: 'If the truth is told, the needs of the natural world are more prosaic than this. A crack in the pavement is all a plant needs to put down roots.'[244] Mabey's unorthodox celebration of (sub)urban nature points to contemporary invocations of the 'new wild' without losing sight of the potentially permanent costs of environmental contamination: 'In areas of gross contamination', he says, 'most plants and animals quite simply die. They have no choice', throwing down the gauntlet in his robust conclusion: 'We do have one, which is to clear up our own filth' (see Chapter 4).[245]

In focusing on Britain's suburbs and industrial hinterlands, Mabey also challenges the view that urban populations are disconnected from nature; on the contrary, he sees city dwellers as playing a significant role in supporting wildlife, especially in the cruel winter months. He recognises at the same time that nature today is more depleted than it has ever been before, acknowledging destructive anthropogenic impacts while taking

[241] Richard Mabey, *The Unofficial Countryside* (Toller Fratrum: Little Toller, 2010 [1973]), p. 20.
[242] *Ibid.*, p. 20. [243] *Ibid.*, p. 19. [244] *Ibid.*, p. 19. [245] *Ibid.*, p. 22.

mischievous pleasure in championing invasive species – from grey squirrels to Japanese knotweed – that are most often regarded as pests. Cities are cosmopolitan by nature, he firmly believes, and there is always a place for newcomers. Such urban post-pastoral (Terry Gifford's term) offers an implicit riposte to the strains of conservatism and xenophobia that are still regularly associated with the rural tradition.[246] It also allows Mabey to discover beauty in the unlikeliest of places, as when he spies a glorious stand of spotted orchids growing out of an abandoned car; or when, in a kind of twilight epiphany, staged at a sewage farm no less, he glimpses wading birds in the fading light, which 'seemed to be appearing from nowhere, limbering up for the next leg of their migration flights through the night skies'.[247]

One of these birds, the dusky redshank (now known as the spotted redshank), is a particular rarity: 'Only a handful are seen every autumn on their long flights from inside the Arctic circle to the Mediterranean.'[248] However, as is often case in the text, the moment of fulfilment is fleeting. As the redshank takes off into the gloom, it is followed by a

> relentless stream of jetliners, bound perhaps for the same destination. They made a glib and obvious symbol, this pair, but one impossible to ignore: the brutish, ear-splitting airliner, burning tons of our dwindling energy resources and spewing its exhaust residue in a sticky film not only over all the plant life in the area, but even, cannibalistically, over other ground-based vehicles; and this swift 12-inch-long bird, whose night-flight noise has caused no more disturbance than to make a few sharp-eared ornithologists nostalgic for lonelier places.[249]

This uncanny pairing speaks to the fragility of nonhuman nature in the face of anthropogenic destruction. Bittersweet images like these position Mabey at the cusp of a new kind of nature writing, paving the way for a concept of the 'environment' that implies a far more entangled sense of human/nonhuman coexistence.[250] To the urban dwellers who, up until this time, had been consuming nature writing of the rural, Mabey revealed that nature was literally on their doorstep – an understanding that pervades much of the popular nature writing of the current moment. That the revelation was double-edged has been confirmed in recent reflections on his own writing. 'Since *The Unofficial Countryside* was first published in

246 See Gifford, *Pastoral*, chapter 7. 247 Mabey, *Unofficial Countryside*, pp. 183, 143–4.
248 *Ibid.*, p. 144. 249 *Ibid.*, pp. 144–5.
250 For this shift in terminology, see Terry Gifford, 'From Countryside to Environment: Reaching Common Ground', in Kate McLoughlin (ed.), *British Literature in Transition, 1960–1980: Flower Power* (Cambridge: Cambridge University Press, 2019), pp. 152–66 (p. 165).

1973, there have been massive changes in the wild land around London', he avers.[251] Gravel pits have been turned into marinas, 'scrubby bomb-sites' into national car parks, and 'the Olympic project has triumphed over the most characterful tangle of wet and wilderness in the East End'.[252] Here, Mabey's verbal echo of Gerard Manley Hopkins' poem 'Inversnaid' adopts a partly elegiac tone: the last thing we humans are able to achieve, he seems to be suggesting, is for wildness and wilderness, wet and weeds to 'be left', as Hopkins asks. Thus, while Mabey may express repeated amazement in nature's adaptability, we are never far in his work, as in the work of the so-called new nature writers who have followed him, from a recognition of acute environmental risk (see Chapter 4). 'There is a limit', he grimly acknowledges, 'to the recuperative powers of the natural world. It will always fight back, but it cannot survive outright eradication.'[253] In this he voices a fear, never far beneath the surface of twentieth-century rural writing, that has only become more urgent in our contemporary times.

[251] Mabey, *Unofficial Countryside*, p. 15. [252] *Ibid.*, p. 15. [253] *Ibid.*, pp. 181–2.

CHAPTER 4

Contemporary

Introduction: Nature Writing after Nature

British perceptions of nature and the natural world in the contemporary era have largely been shaped by television and other media platforms, whether this is on a charmingly local level (*Springwatch*, *Countryfile*) or a spectacularly global plane (*Blue Planet I* and *II*). In a 2005 *Guardian* article that seeks to account for the popularity of *Springwatch*, Blake Morrison argues that 'the urge behind it is an old one: living in harmony with nature', going on to suggest that the show's enduring success is a sign that 'the British public is falling in love with nature all over again'.[1] Whether this is true or not is impossible to prove, but nature programming remains as popular as ever in the British Isles, while its literary cousin, nature writing, is going through a boom phase. Why this is so will be one of the principal subjects of this chapter. To some extent, we will suggest, it is the result of a significant generational shift in public consciousness of the environment, which has given rise to a recent wave of nature writing that responds to the local and global environmental crises which (as argued in previous chapters) are deleterious by-products of the modernity with which they are inextricably entwined.

Just how 'new' this writing is, though, is unclear. Over the past decade or so, various publicity-attracting attempts have been made to make the case for the *new nature writing*, a loose umbrella term under which a number of contemporary writers – several of them doubling as critics – have uneasily sheltered, and sometimes fractious debates around whose provenance and validity have generated a good deal of heat, but precious little light.[2] This is not to say that the case for the new nature writing can't be made, for

[1] Blake Morrison, quoted in Robert Macfarlane, 'Only Connect', *Guardian*, 26 March 2005, n.p.

[2] See, for example, Mark Cocker, 'Death of the Naturalist: Why Is the "New Nature Writing" so Tame?' *New Statesman*, 17 June 2015; Jason Cowley (ed.), *The New Nature Writing* (London: Granta, 2008); Robert Macfarlane, 'Why We Need Nature Writing', *New Statesman*, 2 September 2015; Anna Stenning and Terry Gifford (eds.), 'New Nature Writing', special issue of *Green Letters: Studies in Ecocriticism*, 17 (2013).

commonalities exist within a number of contemporary texts: a renewed tendency to focus on the local and the particular, combined with an up-to-the-minute awareness of the global circumstances that produce them; a registered concern, not just for specific forms of environmental degradation but for the larger social and economic structures within which these are enmeshed; a wistful realisation that no natural landscape or ecosystem today remains unchanged by human interference; and, perhaps above all, a broad ecological understanding of the interconnectedness of human and nonhuman life.[3] Such combined understandings and concerns have fed an environmentally oriented mode of writing that is predominantly written in the first-person singular, but sometimes plays with other voices and perspectives, and characteristically moves across a wide range of different genres, conventions and forms.[4] Such writing is resolutely materialist in its situated attention to the dynamic entanglement of both specific life forms and general earth systems, but it is also residually idealist in its belief in the power of the creative imagination to counteract 'the social and ecological brokenness of our world' and, in so doing, to 'conjoin [environmental] concern with the flourishing of all life'.[5]

If this sounds less like a potted summary of contemporary British nature writing than a mini-history of British nature writing per se, that's probably because it is. In that respect, the 'new nature writing' isn't new at all, but is paradoxically *belated* with respect to its historical precursors. In this chapter, we want to test three hypotheses that further examine this axiomatic condition of belatedness. The first of these revolves around what Edward Said calls *late style*. For Said, 'lateness' is about 'being at the end, fully conscious, full of memory, and very (even preternaturally) aware of the present'; it is also about living out a firmly unreconciled relationship between the present and the past.[6] Lateness and the particular style to which it gives rise are born out of the recognition that there is no state *beyond* lateness; for as Said suggests, following his mentor Theodor Adorno, 'one cannot transcend or lift oneself out of lateness, but can

[3] For different perspectives on the new nature writing, see Deborah Lilley, 'New British Nature Writing', April 2017, www.oxfordhandbooks.com/view/10.1093/oxfordhb/9780199935338.001.0001/oxfordhb-9780199935338-e-155 (accessed 4 February 2021); Joe Moran, 'A Cultural History of the New Nature Writing', *Literature & History*, 23 (2014), 49–63; Kate Oakley, Jonathan Ward and Ian Christie, 'Engaging the Imagination: "New Nature Writing", Collective Politics and the Environmental Crisis', *Environmental Values*, 27 (2018), n.p.; Jos Smith, *The New Nature Writing: Rethinking the Literature of Place* (London: Bloomsbury, 2017).

[4] Jonathan Watts, 'The New Nature Writing: Defining a Genre', *Sounding East*, 5 November 2012, n.p.

[5] Kate Rigby, 'Writing after Nature', *Australian Humanities Review*, 64 (2014), p. 12.

[6] Edward W. Said, *On Late Style* (London: Bloomsbury, 2006), p. 14.

only deepen the lateness. There is no transcendence or unity [to be won].'[7] Lateness emerges out of the fear of being out of *time* (i.e. too late), but also being *out* of time (i.e. untimely or anachronistic). Our contention here is that contemporary British nature writing, for reasons we will go on to explore, is late in both of these primary senses of the term, and the dual consequence of this is that (1) it is haunted by a feeling of inadequacy or impossibility in relation to its predecessors, and (2) it demonstrates some of the frustrations (the lack of reconciliation, the failure of transcendence) that Said identifies in late style.

More specifically, we want to suggest that contemporary British nature writing exhibits some of the anxieties associated with *late modernism*. Jed Esty has argued that late modernism – which he identifies first and foremost as a product of the interwar years – named the response to a perception of aesthetic decline as well as national insignificance, and that this response took the form of trying, if largely failing, to 'reconcile the heterogeneity of [English] culture with the abstract unity of the nation'.[8] One result of this was the mid-century emergence in literature and the arts of a series of 'Romantic moderns', Alexandra Harris' felicitous term for that coterie of English writers and artists in the 1930s and 1940s who were committed to innovation but still passionately interested in the retrieval of the past.[9] For Harris, this gave rise to (among several other things) a bespoke form of English country writing; this in turn supplied the vehicle for a distinctive brand of English rural nostalgia played out in, to give just one prominent example, H. J. Massingham's oeuvre.[10] More recent Romantic moderns such as Ronald Blythe (author of *Akenfield*) and John Lewis-Stempel (author of *Meadowland*) also operate in this pastoral mould, but their renditions of the local are less likely to be old-fashioned than untimely.[11] The titular village of Blythe's study, for example, is temporally suspended between a past it does not especially want and a future it cannot quite bring itself to countenance. Caught in the

7 *Ibid.*, p. 14.

8 Jed Esty, *A Shrinking Island: Modernism As National Culture in England* (Princeton, NJ: Princeton University Press, 2003), p. 185. See also Chapter 3.

9 Alexandra Harris, *Romantic Moderns: English Writers, Artists and the Imagination from Virginia Woolf to John Piper* (London: Thames and Hudson, 2015). See also Chapter 3.

10 Massingham's work is neglected today, but he was one of the most popular rural writers of his time, publishing a string of books during the first half of the twentieth century, including *World without End* (London: Cobden and Sanderson, 1932) and *An Englishman's Year* (London: Collins, 1949). A suitably titled anthology of his writings, *A Mirror of England*, appeared in 1988 with Green Press.

11 Blythe's *Akenfield: A Portrait of an English Village* was originally published in 1969 and has been in print ever since; Lewis-Stempel's more recent *Meadowland: The Private Life of an English Field* (London: Black Swan) has proved equally popular since its publication in 2015.

crossfire of social and economic forces over which it has little control, Akenfield becomes an ideal case study for Blythe's wry examination of the conflict between 'modernity's homesick longing for authenticity and its dogged pursuit of critique'.[12] *Meadowland* is similarly caught between its farmer-author's wish to maintain rural continuity (e.g. through the celebratory replaying of an 'eternal' seasonal cycle) and his partly suppressed awareness of historical changes wrought upon the English countryside which, when seen together, create the forlorn impression of a 'dying [traditional] world'.[13]

Conflicts such as these, argue Tim Milnes and Kerry Sinanan, are at the heart of modernity's ongoing struggle to 'understand itself as self-inaugurating';[14] they are also at the core of modern processes of self-actualisation which instantiate the attempt to retrieve those 'authentic' forms of lived experience that have been marginalised by dominant regimes of technocratic or mechanistic control.[15] As the philosopher Charles Taylor further suggests, this anti-mechanism has its roots in the very Romanticism that self-designated Moderns chose to reject, bringing Romantics and Moderns together in joint protest 'against [those] disengaged and instrumental modes of thought and action which have steadily increased their hold on modern life'.[16]

Contemporary British nature writing offers a continuation of rather than a break from several stereotypically Romantic predilections: fantasies of purity and innocence; dreams and visions; the limits of rationalism and the liberating powers of the creative imagination; revelations of the reality, or perhaps better realities, that lurk behind and beyond visible things.[17] But these proclivities are measured at the same time against the modern rationalist need to explain, demystify and categorise: a need which, while probably finding its apogee in the Victorian period, is deeply embedded within Romanticism itself. What often results in contemporary nature writing is a deep-seated *uncertainty* that expresses itself in both aesthetic and ethical terms, yoking modern existential unease to an updated form of Keatsian negative capability. This uncertainty is *post-Romantic* insofar as it confirms the persistence of Romanticism in the present; but it might

12 Tim Milnes and Kerry Sinanan (eds.), *Romanticism, Sincerity and Authenticity* (London: Palgrave, 2010), p. 17.

13 Lewis-Stempel, *Meadowland*, p. 261.

14 Milnes and Sinanan, *Romanticism*, p. 2.

15 Charles Taylor, *Sources of the Self: The Making of the Modern Identity* (Cambridge, MA: Harvard University Press, 1989), p. 46.

16 *Ibid.*, p. 495; see also Robert Hughes, *The Shock of the New* (New York: Knopf, 1981).

17 *Ibid.*, p. 304.

equally well be considered *late modern* insofar as it occupies a space of hiatus or transition in which lateness becomes an inescapable condition – an incurable affliction that leaves the individual both 'fully conscious [and] full of memory' (Said), precariously poised between the future and the past.[18]

This condition of untimeliness is integral to our second hypothesis, which is that contemporary British nature writing is constitutively belated. We should make clear here that 'belated' is not meant in a postmodernist sense, for whatever else it might be, the 'new nature writing' is emphatically not postmodern. To some extent, it could be argued that recent nature writing in Britain has turned its back on its modernist predecessors, preferring to align itself with nineteenth- or even eighteenth-century traditions of pastoral writing, while grand stylistic experiments in the manner of *The Peregrine* and *The Living Mountain* are a relative rarity (as is the literary excellence of these works).[19] Nor, for that matter, is 'belated' intended in the tormented environmentalist sense conveyed by familiar 'end of nature' arguments. Such arguments, although they take several different forms, usually refer not so much to the wholesale destruction of the environment as to the seemingly irreversible demise of an idea of nature as stable, reassuring and independent: as the American environmental writer and campaigner Bill McKibben puts it, 'nature's independence *is* its meaning, [and] without it there is nothing but us'.[20] Most contemporary British nature writers, one suspects, would take strong issue with McKibben's view that 'the comfort we need is inhuman'.[21] Instead, as previously mentioned, their writing generates much of its momentum from broadly ecological understandings of the mutual entanglement of the human and the nonhuman and the profoundly cultural attitudes and assumptions embedded in the natural world.

Rather, in keeping with what might loosely be called the ecological turn in environmental theory, 'belated' signals the *impossibility of 'getting back to nature'*. This once again aligns modern nature writing with Romantic sensibilities. As the eco-philosopher Timothy Morton puts it, 'nature writing is Romantic insofar as it tries to "get back to nature", and knows

[18] Said, *Late Style*, p. 14.

[19] J. A. Baker's *The Peregrine*, first published in 1967, and Nan Shepherd's *The Living Mountain*, not published until 1977 though written in the 1940s, are widely considered two of the finest 'nature books' of the twentieth century. We will look in more detail at *The Peregrine* later in this chapter; for a detailed analysis of *The Living Mountain*, see Chapter 3.

[20] Bill McKibben, *The End of Nature* (London: Penguin, 1990), p. 54. [21] *Ibid.*, p. 196.

that this possibility is forever excluded'.[22] As he goes on provocatively to suggest, nature writing is an accursed genre, given repeatedly to register the impossibility of its own promises. It counteracts this in part by performing the magic trick of 'ecomimesis', by which it maintains, but also paradoxically dispels, the illusion of immediacy and the impression of being direct.[23]

One need not subscribe to Morton's views to recognise that contemporary British nature writing is belated, both with respect to the earlier literary traditions it rehearses and to those natural processes that, now irrevocably altered under straitened environmental circumstances, it nonetheless replays and re-presents. Three aspects of this belatedness can briefly be mentioned here. At one level, contemporary British nature writing is caught up in literary *postmemory*: that congeries of connected processes by which contemporary writers attempt, through a surfeit of mnemonic representations, to compensate for their inability to gain direct access to the past. At another level, it is inescapably *post-pastoral*. It is bound up, that is, in an ecological awareness of the porous boundary between 'inner' and 'outer' worlds and the renewed environmentalist responsibilities that come with that awareness – the marrying of ecological consciousness and environmental conscience, as Terry Gifford suggests.[24] And at still another level, it is *post-primitivist*, both in the sense that it is unable to resolve its own counteracting impulses towards pastoral retreat and primitive engagement, and to the degree that it is ironically disabused of the same primitive longings it embraces and plays out.

In the next section of this chapter, we will show how some of these longings are crystallised in conflicting conceptions of *wildness* and the *wild*, both long since key categories for Western nature writing (albeit in Britain in rather different ways than in the United States). For now, let us introduce our third hypothesis, which is closely connected to the *Anthropocene*, probably the key term for understanding what many ecocritics, and an increasing number of nature writers, are now referring to as a 'post-natural world'. What does it mean to write 'after nature'? In one respect, all nature writing is after nature insofar as 'the natural world to which it refers is presumed to pre-exist the written text'.[25] However, this argument can be turned around quite easily; as Kate Rigby points out, 'an abstract concept of "nature", at least as it has been articulated in the

[22] Timothy Morton, *Ecology without Nature: Rethinking Environmental Aesthetics* (Cambridge, MA: Harvard University Press, 2009), p. 124.

[23] *Ibid.*, p. 125. For a discussion of ecomimesis in the Romantic period, see Chapter 1.

[24] Terry Gifford, *Pastoral* (London: Routledge, 1999). See also Chapter 3.

[25] Rigby, 'Writing after Nature', p. 1.

Western tradition ... only emerges within the world of full alphabetical writing, beginning with classical Greek'.[26]

Like Rigby, we believe there are sound moral as well as intellectual reasons for writing to be reinstated as secondary in relation to what she calls 'the vastly longer and more complex history of the unfolding of life on earth'.[27] But that doesn't necessarily help us with the perceptual shift from 'after *nature*', the mimetic relationship to nature that Max Horkheimer and Theodor Adorno see as characterising the 'pre-civilised period', to '*after* nature', the rational assertion of human beings over and against nature in what they call the 'historical phase'.[28] Ironically, much contemporary theorising around the Anthropocene – the so-called Human Age marked by the simultaneous ubiquity and intensity of human influence on the planet – has unwittingly reinforced the idea of human centrality even as its ecological underpinnings maintain just the opposite: that human beings, who are as much shaped by as shaping the different worlds they fashion, are co-dependent inhabitants of a precariously shared earth.

This presents – to say the least – a huge problem to both nature writing and the criticism that seeks to account for it. An extreme response to this problem is to claim that the 'traditional' (i.e. post-Romantic) language of nature description is no longer fit for purpose, relying as it does on ideas of place, locality and community that are hopelessly out of date.[29] Another response, equally hyperbolic, is to say that 'nature' itself is no longer a viable term, depending as *it* does on the supposed separation of human beings from the various environments that surround them; indeed, the word 'nature' is an obstacle to rather than a catalyst for environmental action and social reform.[30]

Such views, if universally upheld, would risk putting nature writers out of business. To their likely relief, there are more measured assessments, such as that of the American environmental lawyer Jedediah Purdy, who sees the Anthropocene as the basis for a 'new politics of nature, a politics more encompassing and imaginative than what we have come to know as environmentalism'.[31] Purdy takes care to distinguish between what he calls

26 *Ibid.*, p. 7. 27 *Ibid.*, p. 8.

28 Theodor Adorno and Max Horkheimer, *Dialectic of Enlightenment* (New York: Herder and Herder, 1972), p. 180.

29 See, for example, Timothy Clark, 'Deconstruction, Environmental Politics and the Closure of Ecocriticism', *Oxford Literary Review*, 32 (2010), 131–49.

30 Morton, *Ecology without Nature*.

31 Jedediah Purdy, *After Nature: A Politics for the Anthropocene* (Cambridge, MA: Harvard University Press, 2015), p. 17.

a 'neoliberal Anthropocene', which is marked by the 'ever-intensified management of the globe, carried forward by market means',[32] and a 'democratic Anthropocene', which is morally inclusive, attentive to the equitable distribution of resources, and makes space 'in the imagination and sympathy of people for the nonhuman world'.[33] However, Purdy also points out that the Anthropocene is characterised by an atmosphere of potentially debilitating uncertainty: over what global environmental action to take, over what local social practices and policies to implement, and not least over how to rethink the ways in which both human and nonhuman worlds are imagined within the overall framework of what he ambitiously calls an 'aesthetics of [global] change'.[34]

Such a transformative aesthetics, he suggests, urgently needs to finds a place within it for the *uncanny*: for the 'unsettling perception that we do not know, perhaps cannot know, the ethical status, meaning, or experience of another living being that stands in front of us'.[35] Uncanniness, for Purdy, by no means entails a rejection of the beautiful and the sublime – those twin pillars of conventional environmental aesthetics – but whereas the beautiful and the sublime are based on what we already know, the uncanny confronts us with our lack of knowledge: with the gap between our *appreciation* and our *understanding* of what we apprehend. Uncanniness, he says, focusing on the human-animal domain, begins 'from admitting that we don't know what an animal's life means, to it or us, and that the answer we give can only be some blend of discernment and projection'.[36] This then extends to perceptions of the environment, which 'now that all simple ideas of nature are irretrievably gone' requires different modes of cognitive and emotional engagement, such as 'learning to find beauty in ordinary places . . . or to prize what is neither pure nor natural but just is – the always imperfect joint product of human powers and the natural world'.[37]

We have dwelt on Purdy here because his work, US-centred though it is, goes further than most in staking out the territory for contemporary UK-based nature writing, which as we will see in this chapter brings together just the combination of characteristics (uncanny aesthetics, epistemological uncertainty, ecological interconnectedness) he explores. Purdy's emphasis on *failure* – the unbridgeable 'disconnect' between experience and representation, the gulf of incomprehension between different life-worlds, the ongoing failure that is environmental politics[38] – similarly looms large in contemporary British nature writing. Such failures, which are by no means

[32] *Ibid.*, p. 48. [33] *Ibid.*, p. 50. [34] *Ibid.*, p. 245. [35] *Ibid.*, p. 230. [36] *Ibid.*, p. 244.
[37] *Ibid.*, pp. 9–10. [38] *Ibid.*, p. 255.

new, are inherited from what Purdy calls a 'long and complicated legacy of environmental imagination and practice'.[39] In the British case, they are also tied in with the specifics of *national* heritage: destructive practices of enclosure and deforestation; exploitative histories of agricultural reform; the compensatory shoring up of an official image of the (usually English) countryside that is both a reflection of the multidimensional losses entailed by these practices and an imaginative salve to them (as in the turning of English rural history itself into a commercially viable heritage form).

What *is* new, perhaps, is the degree to which contemporary nature writers are prepared to engage with them. British nature writing is nothing if not a reflexive form, and as has been argued throughout this book its reflexivity owes a lot to the always volatile, sometimes violent, modern circumstances and conditions (industrialisation, urbanisation, the shrinkage of the commons, and so on) that are the paradoxical grounds of its existence. What we find, then, is not so much the 'end' or 'death' of nature, as nature wilfully controlled and domesticated, nature opportunistically exploited, nature anything but tranquil in anything other than imaginatively recollected form. It is not so much that contemporary British nature writing trails in the wake of this far-from-consolatory tradition as that it intensifies it. For all its occasional veneer of pastoral contentment, conflict and uncertainty are at this writing's very heart, and, always aware of the impossibility of the tasks it sets itself, it remains ironically attuned to – if not necessarily dismissive of – the various restorative ruses it strategically deploys.[40]

Recalling the Wild

'The wild' is a difficult term and, as might be expected given its different cultural inflections, there is little critical agreement about what it signifies. For the American poet and critic Gary Snyder, wildness is not just a particular quality that can be ascribed to ways of experiencing the

[39] *Ibid.*, p. 9.

[40] Such ruses may include versions of pastoral retreat, the attempted recovery of personal and/or collective memories and the restorative 're-enchantment' of landscape itself. An important nature-writing form that can combine all three of these elements is the so-called recovery narrative, contemporary examples of which include Richard Mabey's *Nature Cure* (2005), Helen Macdonald's *H Is for Hawk* (2014) and Amy Liptrot's *The Outrun* (2016). As might be foreseen, contemporary narratives tend to be sceptical towards the curative attributes of the natural world even as they invoke them. However, it is quite possible that nature writing continues to be *read* in therapeutic and/or reassuring ways that the writers themselves might wish to challenge. More work urgently needs to be done in this area, which is beyond the remit of this book.

world; it is the world itself, which offers a place or, perhaps better, a myriad of possible places where the wild potential of all living beings may be examined and expressed. Wildness, in this sense, is ubiquitous; unlike wilderness, which may erode, wildness endures, and can be found at all levels, both biotic and abiotic: in 'the wind, the desert sands, [or] the millions of tiny seeds of original vegetation [that are] hiding in the mud on the foot of an arctic tern'.[41] Wildness is also within, and what Snyder seeks is a 'bonding of the wild [within] ourselves to [the wild] process of the universe'.[42] For this to happen, nature must be 'admitted from within, as a quality intrinsic to who we are';[43] it cannot be objectified. Snyder is duly ambivalent about nature writing, which he sees as a primarily descriptive mode that depends too much on the 'habits of collection and classification' that are the accompaniments of civilised life.[44]

Notwithstanding, wildness has become the standard currency of Western nature writing, even as its meanings and locations have shifted significantly over time. As might be expected, British nature writers have tended to take a rather different view of wildness to that of their American counterparts, for in the 'semi-natural' conditions that have obtained in Britain since the Middle Ages, wildness – in whatever simple guise or complex manifestation – has often been difficult to find. Difficult, but by no means impossible, and even today there are pockets of wilderness in the British Isles that have proved magnetically attractive to nature writers. More evident still have been recent critical attempts to redefine wildness – always a relational term – in the scrambled context of the Anthropocene, which encompasses the everyday entanglements of human and nonhuman life. For the ecocritic Jos Smith, 'wild' functions as an indeterminate zone that operates 'at the human/non-human interface';[45] for the geographer Jamie Lorimer, it is less an intensification than a paring down of bodily experience, which is further identified with the typification of place (mountain, island, valley, etc.).[46] For Smith and Lorimer (and several others like them), *wildness* is best uncoupled from *wilderness* – which continues to be associated with the remote and unpopulated – and aligned instead with everyday experience in a 'hybrid world' that is neither social nor natural, but both of these things, and in multiple possible variations, at once.[47]

[41] Gary Snyder, *The Practice of the Wild* (Berkeley, CA: Counterpoint Press, 2010), p. 12.
[42] *Ibid.*, p. 16. [43] *Ibid.*, p. x. [44] *Ibid.*, p. 194. [45] Smith, *New Nature Writing*, n.p.
[46] Jamie Lorimer, *Wildlife in the Anthropocene* (Minneapolis: University of Minnesota Press, 2015), p. 11.
[47] *Ibid.*, p. 12.

Similarly, and in keeping with its ecological remit, contemporary British nature writing describes an inherently messy world in which 'pure' experience is inaccessible, and where wildness often reverts to its more negative association with confusion, even violence, and unruly behaviour of different kinds. Such forms of wildness recall primordial states, but nature writing (whether 'new' or not) recognises its own incapacity to restore these; as the American ecocritic Lawrence Buell puts it, it practises a 'conceptual restorationism that reorients the denaturized reader, not to a primordial nature, which we cannot recover either in fact or fantasy, but to an artifactual version of environment [that is principally] designed to evoke place-sense'.[48] This view of nature writing (again, whether 'new' or not) is charitable to the extent that it implicitly accepts the genre's persistent impulse to compensatory nostalgia – its residual pastoral leanings – at the same time as it invokes a post-primitivist primordialism that recalls the wild even though it knows that primordial wildness is out of reach.

Such performative primitivism is more in evidence in American than in British nature writing: the part effect of a 'place-sense' that continues to be fashioned very differently in the two countries. In the United States, the trope of *wilderness* has been key in registering a particular form of American exceptionalism that opposes 'New World' to 'Old World' experience. One version of this is the cultural-nationalist view that wilderness constitutes a precious 'American asset, [both] a cultural and moral resource and a basis for national self-esteem';[49] another is the more challenging idea that 'there is almost no real wilderness left [in the British Isles] and absolutely no memory of it . . . The English, the Europeans, are too far from the wild. That's the difference between them and us.'[50]

Such blanket distinctions are primarily ideological. They also tend to mistake wilderness for wildness, assuming that the geographical attributes of the one can be mapped on to the moral qualities of the other, and that, in Roderick Nash's more inclusive language, 'wilderness [represents a vast] reservoir of wildness vitally important for keeping the spirit of wildness alive in man'.[51] Contemporary British nature writing offers a strong ideological countercurrent to the American wilderness ethic. It refutes

48 Lawrence Buell, *The Environmental Imagination: Thoreau, Nature Writing, and the Formation of American Culture* (Cambridge, MA: Harvard University Press, 1995), p. 267.
49 Roderick Nash, *Wilderness and the American Mind* (New Haven, CT: Yale University Press, 1982), p. 67.
50 William Least Heat-Moon, quoted in Robert Macfarlane, *The Wild Places* (London: Penguin, 2008), p. 9.
51 Nash, *Wilderness and the American Mind*, p. 88.

the view that there are no wild places left in the British Isles; it also conjures up distant memories of wilderness. Perhaps more than either of these, though, it plays between alternative ideas of wildness as these have moved through time; indeed, for many of the so-called new nature writers, wildness is as much a *temporal* as a spatial category, and their engagement with the wild is not so much an act of remembrance as a telescoping of human and nonhuman history that allows for an exploration of the recesses of deep time.

Paramount among these is the Cambridge-based writer-academic Robert Macfarlane, who has probably done more than anyone to make a case for the 'new nature writing' as a distinct British enterprise. It seems reasonable to start, then, with an analysis of Macfarlane's most detailed examination of the wild to date, his celebrated 2007 travelogue *The Wild Places*, which we will read in terms of both the physical geography and imaginative archaeology of wildness: as a multifaceted attempt at salvage in which the past is not necessarily recoverable and the histories it discloses are never pristine or pure. We will go on to look at two contemporary British works which, even if they fit less well than Macfarlane's into standard definitions of the new nature writing, still revolve around similarly conflicted ideas of wildness: Jay Griffiths' *Wild: An Elemental Journey* (2006), and George Monbiot's *Feral: Searching for Enchantment on the Frontiers of Rewilding* (2013). We will then close the section with some thoughts on a recent text that engages more specifically with the 'new wild', Paul Farley and Michael Symmons Roberts' *Edgelands: Journeys into England's True Wilderness* (2012), which distances itself from an older, allegedly misanthropic tradition of wilderness writing while reclaiming a series of semi-abandoned places as uncanny sites for a 'true pastoral' that revolves around shifting registers of time and space.[52]

***

A brilliantly realised twenty-first-century version of the medieval wonder voyage, Robert Macfarlane's contemporary nature odyssey *The Wild Places* sets out to rediscover wildness in a series of affectionately rendered British landscapes which together form a kind of memory map that links multiple histories to one another: histories of the planet, histories of the nation, histories of the self. The map constitutes less a rational grid than an emotional cartography that attaches feelings to stories and vice versa, seeking out 'lost way[s] of proceeding within a landscape' that shuttle

[52] Paul Farley and Michael Symmons Roberts, *Edgelands: Journeys into England's True Wilderness* (London: Penguin, 2012), pp. 166, 105.

between human memory and natural form.[53] As befits this cartography, the structure of the text is both linear and recursive, moving from one landscape to another but returning insistently to the feelings and memories that link them within a 'web of wild places' that stretches beyond the shores of Great Britain to take in the wider world (p. 292). At other times, the web contracts, framing more familiar places, experiences and acquaintances, and rediscovering the wild in the familiar, in what Macfarlane calls 'the little places' (p. 237): backyards, roadsides, hedges; the domestic world of 'semi-nature' (p. 7) in which wildness is replayed in a minor key.

As suggested earlier in this chapter, *The Wild Places* is written against the view that there are no wild places left in the British Isles, and that wildness must be associated with wilderness, with Romantic notions of 'wide spaces, [figureless and] remote' (p. 7). It is also written against the idea that the only tradition that counts in British nature writing is that of pastoral elegy, in which a heavily reworked national landscape only occasionally discloses 'suppressed memories of a more ancient archipelago', and wild places are no longer marked, even if they have not definitively disappeared (p. 10). Instead, Macfarlane's task is to bring these places back to life, to salvage them and uncover their secrets; also to aestheticise them, for his text is the carefully crafted work of a naturalist-connoisseur who is as much interested in the shape and sound of words as in the landscapes they describe. The term 'landscape' itself, with its connotations of embellishment, confirms that there is hardly 'a single . . . natural system that has not, for better or worse, been substantially modified by human culture'; but it also provides a reminder that such systems have provided the raw materials for some of the world's finest art.[54]

The wild, in this context, is very much a *fashioned* idea, a human projection onto the nonhuman world with which it is irrevocably entangled; but as *The Wild Places* shows, it also contains within it an entire *history* of ideas that have 'moved immensely through time' (p. 30). Early on in the text, Macfarlane compares two seminal ideas of wildness, as a 'quality to be vanquished' and as a 'quality to be cherished', connecting both to illustrative stories (p. 30). According to the first story, 'wildness has been perceived as a dangerous force that confounds the order-bringing pursuits of human culture and agriculture . . . Wild places resist conversion to human use, and they must therefore be destroyed and overcome' (p. 30). In the second, by contrast, wildness is 'an energy both exemplary and exquisite, [and] wild places are realms of miracle, diversity and abundance' (p. 31).

[53] Macfarlane, *Wild Places*, pp. 143–4. Further references are in the text.
[54] Simon Schama, *Landscape and Memory* (New York: Knopf, 1995), p. 7.

It seems clear where Macfarlane stands – for what he later calls 'the wildness of natural life . . . vigorous and chaotic' (p. 316) – but it is much less clear, at least initially, how he positions wildness in relation to human time. In this last context, *The Wild Places* might best be described as moving from one thesis on wildness – that 'to reach a wild place [is] to step outside human history' (p. 7) – to another, that 'thousands of years of human living and dying have destroyed the possibility of the pristine wild' (p. 127). The text gravitates towards this second view, which recognises that 'every islet and mountain-top [in the United Kingdom], every secret valley or woodland, has been visited, dwelled in, worked, or marked at some point in the past five millennia' (p. 127). 'The human and the wild can [no longer] be partitioned', Macfarlane concludes – if they ever were (p. 127).

There is more, though, for as Macfarlane acknowledges, 'the wild prefaced us, and it will outlive us' (p. 316): there is thus a different temporal order of wildness that extends beyond the human into the 'shifting sedimentary histories' of geological deep time (p. 217). *The Wild Places* sifts through these histories, some of which are accessible only to the imagination; even travel overland comes to have an oddly subterranean dimension to it, as if the experience of wildness were less a process of immersing oneself in place than of submerging oneself in time (p. 217). As Macfarlane suggests, there is a geological logic – a 'geo-logic' (p. 242) – in his journeys by which the history of Britain and Ireland can be comprehended 'through the history of its six great rock types – granite, sandstone, slate, chalk, limestone and flint' (p. 243). These rock types form the metaphorical backbone, 'the strong mineral skeleton', of the archipelago, and Macfarlane consoles himself with the knowledge that 'whatever we did to the skin of the country, the skeleton would remain' (p. 244).

In keeping with this 'geo-logic', excavation metaphors can be found throughout the text, which also literally goes to ground in a variety of cavities and hollows: the sunken roads (holloways) that 'constitute a labyrinth of wildness in the heart of arable England' (p. 218); the vast natural ossuary that is the Burren in Ireland, its cracked limestone surface riddled with unmarked graves (p. 170). At different levels, and on different timescales, *The Wild Places* thus performs what – adapting James Clifford – we might call a *salvage geography*, which is not so much about rescuing particular places, rather about rescuing a particular way of thinking and feeling about place that has become eroded over time.[55] Salvage geography

[55] See James Clifford, 'On Ethnographic Allegory', in James Clifford and George Marcus (eds.), *Writing Culture: The Poetics and Politics of Ethnography* (Berkeley: University of California Press, 1986), pp. 98–121.

is an act of reclamation that recognises wildness has been lost, but may yet have the potential to reseed itself (p. 281); it is also an act of reconnection that joins place to place and people to each other and the wider natural world (p. 314). At the same time, salvage geography, in attaching memories to place, also recognises that place outstrips the capacities of human memory. Hence Macfarlane's energetic attempts to conjure up an ancient world in the hope of narrowing a gap he fears may be unbridgeable – attempts which join a post-pastoral consciousness of continuing histories of environmental damage to a post-memory sensibility of originary loss.

In this and other ways, Macfarlane's text struggles against its own Romantic impulses, and is repeatedly drawn back to the very histories of violence and exploitation from which it claims to want to escape (p. 7). *Wildness* is the multivalent code word for this struggle, which can never be won, in part because the violence it instantiates is unappeasable; yet this same violence supplies the dynamic link between wildness's 'tutelary harshness' and its 'exuberant vegetable life' (p. 176). Macfarlane thus comes to recognise that there is no essential difference between a 'human' wildness, which exhilarates, and an 'inhuman' one, which annihilates (p. 157). *Both* must be accommodated if wildness is to be appreciated across its different registers as an intensification of experience, an apprehension of the physical and emotional density of the natural world. Similarly, Macfarlane is brought to the realisation that if in wildness is the preservation of the world, it is also a symptom of human destructiveness – hence the circular logic of his closing aphorism that 'we are fallen in mostly broken pieces [but] the wild can still return us to ourselves' (p. 20).[56]

***

[56] A major writer who reflects intensively on this destructiveness is the late W. G. Sebald, whose capacious term the 'natural history of destruction', ostensibly concerned with the carnage of Allied bombing during the Second World War, opens out onto a more general – profoundly melancholic – consideration of historical inevitability: the gravitation of all civilisations, and the individuals and societies caught up within them, towards erosion and loss. While Sebald's work has been hugely influential in the English-speaking world as well as in his native Germany, it is not easy to make a case for him as a 'nature writer' despite his obvious interest in natural history and passages of superb natural description scattered throughout his work. However, there is also a sense in which Sebald's work is *exemplary* of modern British nature writing in its emphasis on multiple crises and its mournful contemplation of the violence human beings have historically inflicted on the natural environment, on each other and on themselves. Sebald's work has been extensively commented upon elsewhere and is regrettably not covered here, but works of his that could easily have merited inclusion in this study are the aforementioned *On the Natural History of Destruction* (London: Penguin, 2004 [1999]), the long poem *After Nature* (London: Penguin, 2003 [1988]) and the part-fictional travelogue *The Rings of Saturn* (London: Vintage, 2020 [1995]).

A very different kind of disquisition on the wild is Jay Griffiths' eponymous *Wild: An Elemental Journey*, which harnesses the anarchic, if not necessarily destructive, attributes of wildness to a deliberately ramshackle travelogue that veers between the Amazon and the Arctic, imitating the four cardinal elements it tracks. Griffiths' text appears initially to have little in common with Macfarlane's, and although she is British by nationality, hers is a travel narrative written mainly in the American transcendentalist tradition of writing about the wild. However, like Macfarlane, Griffiths is interested in the etymology of wildness, its connections to a variety of folk traditions and its popular associations with vitality and freedom, the exhilaration that comes from openly defying convention and taking risks that are potentially life-threatening, but that also serve to confirm the value of life itself. *Wild* operates in this vitalist tradition as an unruly miscellany of self-willed terms, in which 'wild' is less an adjective or a noun than a *verb*: one synonymous with transgressive actions and with those kinds of directionless travel in which vagrancy is a form of deliverance, destination is of no importance, and to follow the open road is to follow what Griffiths alliteratively calls 'the fire-fare-forwards of life itself'.[57]

Throughout *Wild*, women are championed as the ultimate wanderers, although they have historically been vilified for doing so, while their male counterparts are negatively associated with the type of phallic adventurism that fuses masculinity, nationalism, imperialism and military might (p. 345). Griffiths' own adventures are of a different kind, not least because she inhabits a different body, and the mythical figure of the 'wild woman' features frequently as the embodiment of an Amazonian desire for something beyond human society, just as the Amazon itself becomes an objective correlative for Griffiths' own 'personal wilderness, filthy, earthy, jungly, roaring with its waterfalls, its white cascades – a self-willed, claggy swamp' (p. 61). Here as elsewhere, Griffiths seems happy to recycle clichés about women as 'forces of chaos, changeable as the seas, surprising, capricious, fickle' (p. 217). Indeed, a distinctive feature of the text is that it appropriates derogatory female stereotypes and turns them back again into virtues, especially those surrounding 'the wild mobility of women', whose sexual freedom has been reinterpreted as licentiousness, and whose thinking and conversation have been judged as unnecessarily rambling and digressive, as if designed deliberately to infuriate the hyper-rationalist masculine mind (p. 306).

[57] Jay Griffiths, *Wild: An Elemental Journey* (London: Penguin, 2006), p. 132. Further references are in the text.

Griffiths' own language accordingly takes pride in making huge intuitive leaps between different ideas, and much of the text is as unruly as its own subject, simulating a chaotic energy it seems both unable and unwilling to contain. The chapter on 'Wild Water', for instance, moves from maps to whales, then to a short riff on utopia, with oceans – the ostensible link between these – seen as a vast 'theatre of desire' (p. 233). As in other chapters, the pathetic fallacy pervades, so that the entire natural world is seen as an extension of personal feelings, and 'inner' and 'outer' experience are made to mirror one another, any nominal boundary between them definitively undone: '[It] is real liberation to know the wild boundlessness of the oceans within, played out in the oceans of this blue world . . . I know I am oceanic. I fathom it in other women too. I know we can speak at the shoreline and feel it in our depths' (p. 214).

Passages like this one suggest that *Wild* has more in keeping with late twentieth-century Anglo-American ecofeminism than it does with the ecological dimensions of the new nature writing; still, the text's phenomenological emphasis positions nature writing after the so-called affective turn, which explores how emotions are lived and experienced in bodies, and how organic bodies interact with larger systems of non-organic life.[58] Such writing rejects the blandishments of pastoral, preferring instead to focus on a wild vision of nature as confused and disorderly, but also fecund and dynamic, doing as it pleases, refusing to be told (pp. 49–50). For Griffiths, wild nature is free, and it also nurtures other fundamental freedoms: 'freedom of will, freedom of time, and freedom of thought' (p. 87). She sees such freedoms, problematically perhaps, in some of the world's indigenous peoples, whose own fights for another kind of freedom – political freedom – she passionately supports (p. 370). Her political sympathies are with the underdog, and she is drawn throughout her travels to spend time with people whom she sees as living at the 'emotional' centre of a world where others, with greater economic and political power, 'rationally' position them at the edge. She recognises at the same time that such people are very much part of a modern world that depends on them – and on the natural resources that fall within their ancestral heartlands. Thus, while at times *Wild* sounds like Gary Snyder in its universal support for 'nature-not-programmed' (p. 373), at others it

[58] For different perspectives on affect and the affective turn, see, for example, Sara Ahmed, *The Cultural Politics of Emotion* (Edinburgh: Edinburgh University Press, 2014); Stacy Alaimo, *Bodily Nature: Science, Environment, and the Material Self* (Bloomington: Indiana University Press, 2010); Patricia Ticineto Clough and Jean Halley (eds.), *The Affective Turn: Theorizing the Social* (Durham, NC: Duke University Press, 2007).

echoes Rachel Carson in its situated rage at nature desecrated and exploited, like the mountain at Freeport which, sacred to local West Papuan peoples, is used by an American mining company as a convenient dump for toxic sludge (p. 398).

Griffiths' text emerges in contexts like these as an extended diatribe against what she calls 'the taming of the ... self-willed human spirit' (p. 376) – a spirit that resides in all of us, but that few of us have the courage to enact. One term for this untamed spirit is the *primitive*, and *Wild* – at first sight at least – is nothing if not a primitivist text. Almost point for point, Griffiths' text matches Marianna Torgovnick's checklist of popular Western primitivist tropes, echoing their counter-modernist longings for a world in which we are reconciled to 'our id forces [and] our untamed selves'.[59] *Wild* also matches the more gendered aspects of Torgovnick's later work on primitivism, for example, her 1998 study *Primitive Passions*, where she looks at the primitive in terms of the desire to fully inhabit the body, to enjoy the sensual pleasures it offers, but also to transcend bodily limitations in search of a higher self.[60]

In other respects, however, Griffiths' is a post-primitivist text in all but its most illusionary registers. Its post-primitivism shines through in the overriding sense that the 'primitive world' is dead, even if its myths live on in the sometimes pathological desires of moderns to live 'like primitives' (MacCannell), and in the more restricted sense that – as Griffiths painfully witnesses on the shores of the Amazon – 'the last tribal people are [now] living in slums, morosely selling postcards of how they used to live' (p. 95).[61] Some of Griffiths' more riotous invocations of the wild – her drug-addled, pseudo-shamanic vision of herself as a 'jaguar's apprentice' (pp. 98–102), her closing Rabelaisian assertion of the subversive power of 'wild comedy' (pp. 409–21) – need to be balanced against these melancholic insights, which read less like reinvigorated Rabelais than second-hand Lévi-Strauss. And, like Lévi-Strauss, Griffiths locates the wild as a function of a desire that tacitly acknowledges its own self-deceptiveness. In a short section of chapter 2 ('Wild Ice'), Griffiths cleverly juxtaposes Lévi-Strauss (*The Savage Mind*) and Jack London (*The Call of*

59 Marianna Torgovnick, *Gone Primitive: Savage Intellects, Modern Lives* (Chicago, IL: University of Chicago Press, 1990), p. 8. See also the discussion of primitivism, via the work of D. H. Lawrence, in Chapter 3.

60 Marianna Torgovnick, *Primitive Passions: Men, Women, and the Search for Ecstasy* (Chicago, IL: University of Chicago Press, 1998).

61 See also Dean MacCannell, *Empty Meeting Grounds: The Tourist Papers* (New York: Routledge, 1992).

the Wild) to make the case for an originary human wildness.[62] In *The Call of the Wild*, Buck the dog hears wolves calling and is precipitated back to his own pre-domesticated origins. We too hear that call, London suggests, and it lives on in us because we, unlike dogs, have never been domesticated: because wildness is our original state (pp. 149–50). In *The Savage Mind*, similarly, Lévi-Strauss recalls the human mind 'in its untamed state' – that is, prior to society's civilising influence. 'What then is the wild human', he asks rhetorically. 'Is it savages? It is us!' (p. 150). Griffiths appears to miss the point that both London and Lévi-Strauss are being ironic. Original wildness is only accessible in memory. After all, London's wild animal 'remembering' is inevitably projected through *human* memory, while these secondary memories are filtered further through the 'citational memory' (Plett) processes by which Griffiths recalls London's and other previous texts.[63] But perhaps that is Griffiths' claim: that wildness is only ever experienced second-hand, through our apprehension of others' wildness, and through the strategic appropriation of that wildness for ourselves.

***

So far we have been arguing that the 'new nature writing' – insofar as it exists at all – practises what Simon Schama felicitously calls a 'primitivist pastoralism', a hybrid aesthetic mode characterised by its ironic consciousness of belatedness, which recognises the need for human interference in nature but still believes at some deeper level in the self-regulating capacities of the natural world.[64] More specifically, the self-willed character of the natural world is mapped on to the individual that apprehends it, so that wildness becomes both the spectral projection of a repeatedly frustrated desire for personal freedom and a complementary confirmation of the reciprocal, historically shifting connection between nature and culture, world and self.

This dual idea of wildness is central to the work of the contemporary English environmental writer-activist George Monbiot, whose 2013 book *Feral* doubles as an angry denunciation of a 'dewilded' world that severely 'limits the range of [human] engagement with nature, [pushing] us [all] towards a monoculture of the spirit' and a sustained

62 Claude Lévi-Strauss, *The Savage Mind* (Chicago, IL: University of Chicago Press, 1966); Jack London, *The Call of the Wild, White Fang, and Other Stories* (Oxford: Oxford University Press, 2009 [1903].

63 For further reflections on citational memory, see Heinrich Plett, *Intertextuality* (Berlin: Walter de Gruyter, 1991).

64 Schama, *Landscape and Memory*, n.p.

paean to the wild energies of nature and their capacity to reinvigorate human life.[65] *Feral* offers a characteristically impassioned account of 'the rewilding of human life' (p. 10), which Monbiot attaches both to his own personal quest to find a way beyond the deadening experience of 'ecological boredom' (p. 7), and to a collective attempt to 'rewild' natural ecosystems that have become seriously depleted, some of them apparently beyond repair. Rewilding is not the small-r romantic idea of restoring such systems to their putatively original state, which its practitioners recognise as illusory; rather, it refers to the pragmatic need to boost their inbuilt capacity for regeneration, for example, by performing reintroduction experiments that might encourage natural ecological processes to restart (p. 8).

The ecosystems Monbiot primarily has in mind are those of Great Britain, which he dolefully sees as an ecological disaster zone all but ruined by intensive agriculture and other industrial attempts to manage the countryside, and marked by a longer history of environmental destruction in which successive acts of enclosure have led to a drastic shrinkage of the commons and a concomitant diminishment of the imaginative potential of social life (pp. 167–8). For Monbiot, modern conservation strategies are not enough to repair the damage caused by centuries of environmental abuse, which has usually gone hand in hand with the maintenance of (landowning) class interests; on the contrary, conservation is part of the problem insofar as it 'manage[s] nature as if it were tending a garden', and thus ends up by 'freez[ing] living systems in time' (p. 8).[66]

Enter rewilding, which Monbiot ambitiously invests with the collective hopes of a 'positive environmentalism' that 'offers new freedoms in exchange for those we have sought to restrict' (p. 12), and in which less emphasis is placed on protecting things that are already there – which he reductively sees as the primary, curatorial purpose of conservation – than on permissively creating the ecological conditions under which things can come back (p. 224). Like other popular apologists for rewilding, Monbiot is prone to see it as a panacea for disastrous times, though he does well to resist the moral grandstanding of Caroline Fraser, whose 2009 account sees rewilding as nothing less than the main arm of a global 'conservation

[65] George Monbiot, *Feral: Searching for Enchantment on the Frontiers of Rewilding* (London: Allen Lane, 2013), p. 154. Further references are in the text.

[66] For a more detailed discussion of modern conservation and its discontents, see Chapter 3.

revolution' in the context of the Sixth Great Extinction, and who melodramatically envisions it as 'a Marshall Plan for the planet' through which, having previously played the part of destroyers, we humans now have the chance to refashion ourselves as creators, 'recovering wastelands and making them whole'.[67]

Monbiot recognises the absurdity of such claims, and he also sees the pitfalls of a global rewilding philosophy that seeks to impose itself on people or parts of the world that may not necessarily be benefitted by it, and which in some cases may almost certainly be harmed (p. 12). Still, *Feral* offers an idiosyncratic view of rewilding, heavily motivated by its author's curious obsession with British 'zoophobia' (p. 112) – an ascribed national deep fear of wild animals – and his almost visceral, eco-anarchic antipathy for management of any kind. As the scientific literature suggests, rewilding theory is often simplistic in its social analysis and insufficiently attuned to the competing moral obligations it sets in play.[68] At worst, it is also seemingly oblivious to its own managerialism, despite being clearly based on an interventionist approach to nature in which the 'intended re-introduction' of species always risks turning into an 'unintended invasion'[69] – an especially good example, this last, of the contradictions embedded within the culturally constructed category of wildness, which are embarrassingly laid bare when nature really *does* find its own path.[70]

Feral is arguably open to all three charges, especially the third, but to accuse Monbiot of getting rewilding wrong is to overlook its *imaginative* function within what might generally be described as the text's animated archaeology: its multiply replayed attempts to uncover links to prehistoric nature-cultures that are still part of who we are (p. 39). Probably the best examples of this are in the early scenes in which we find Monbiot in Pembrokeshire (Wales), equipped with kayak and spear, almost comically rehearsing the hunting instincts of his Mesolithic

[67] Caroline Fraser, *Rewilding the World: Dispatches from the Conservation Revolution* (New York: Metropolitan Books, 2009), pp. 5, 14, 281.

[68] For a balanced view, see Calum Brown, Robert McMorran and Martin F. Price, 'Rewilding: A New Paradigm for Nature Conservation in Scotland?', *Scottish Geographical Journal*, 127 (2011), 288–314; for a more critical approach, see also John Hintz, 'Some Political Problems for Rewilding', *Ethics, Place and Environment*, 10 (2007), 177–216.

[69] Jelte van Andel and James Aronson (eds.), *Restoration Ecology: The New Frontier* (Oxford: Wiley-Blackwell, 2006), pp. 24–5.

[70] Jamie Lorimer and Clemens Driessen, 'Wild Experiments at the Oostvaardsplassen: Rethinking Environmentalism in the Anthropocene', *Transactions in the Institute of British Geographers*, 39 (2013), 169–81.

forebears, though he claims with all due seriousness that what he is experiencing is 'a genetic memory', and that what he is doing is uncannily like 'something I had done a thousand times before' (pp. 33–4). Monbiot's excursus on rewilding is linked, here as elsewhere, to a wider debate on evolutionary memory in which palaeoecology becomes 'a portal through which we may pass into an enchanted kingdom', and present ecosystems are intuited as 'the spectral relics of another age, which, in evolution's timescale, is still close [to our own]' (p. 93).

As in *The Wild Places*, the search for wildness morphs into an imaginative exercise in deep time in which individual and ecological revival – the twin purposes of rewilding in *Feral* – are experienced in ancestral terms (p. 255). At the same time, Monbiot recognises that thought experiments of this kind are exactly that and that there is something irresponsible about the whole idea of an epic vision, replete with 'struggles with the beasts of prehistory' (p. 139), which looks to clear imaginative space even as the real spaces of the present are increasingly circumscribed and constrained. Hence his scepticism towards surely one of the oddest contemporary experiments in rewilding, Pleistocene Park in Siberia, where a plan is apparently being hatched for the reintroduction of the woolly mammoth, which is envisioned, *Jurassic Park*–style, as being resurrected by injecting genetic material from frozen corpses into Asian elephant eggs (p. 142). Such sensationalist ventures may light imaginative fires, but they are difficult to justify. They also draw attention to the moral vacuity of reintroducing 'exciting' species (lions, elephants, and other so-called charismatic megafauna) irrespective of whether or not they are ecologically well matched (pp. 139–40).

There is thus a sense in which Monbiot is slightly disabused of his own imaginative project to rewild the world, although it is clear that he stands by his more grounded 'exploration of nature's capacity to regenerate itself' (p. 268), and that he truly believes that his witnessing of 'the potential for wildlife to return to the places from which it [has] been purged' has enriched his own personal life (p. 268). Whether *Feral* provides a more general antidote to ecological boredom is another matter: Monbiot's text is too self-centred, too thrill seeking and too concerned with its own unfulfilled yearnings for 'improbable glory' (p. 259) for that. Monbiot's wild is, precisely, *his* wild, as it has been for generations of British nature writers, and in fact it is this insistence on the validity of personal experience that helps to make it a highly effective piece of nature writing, if not a particularly convincing environmental text. As with so many examples of the genre, both 'new' and not, the writer's 'dual accountability' – to the object world and to the imagination – is folded into a greater accountability to the self.[71]

***

The wild means different things to different people: a quality of self, a relation to the world, an atavistic memory. Contemporary British nature writing, especially those environmentally conscious forms that have been associated with the new nature writing, explores the broken connections between these, reinterpreting them in the light of a global environmental crisis characterised by a destructively changing climate and by devastating species loss. In its more specifically British guise, it also reinterprets the national narrative, revisiting the primal violence that accompanies it and challenging the pastoral images behind which that violence hides. To recall the wild is to expose oneself to such violence and to the conflicted histories that lie behind it; for far from glorifying an imaginary 'Old England', 'a landscape built from words, woodcuts, films, paintings, picturesque engravings', the wild offers an unwanted reminder of the violent underside of English pastoral nostalgia – of the bloody wars fought to create and protect often wrongheaded ideas of national belonging and bucolic calm.[72] As the English nature writer Helen Macdonald – whose work will be analysed in more detail later in this chapter – puts it, 'I wish that we [English] would not fight for landscapes that remind us of who we think we are. I wish we would fight, instead, for landscapes buzzing and glowing with life.'[73]

Such landscapes aren't necessarily the ones we might expect to find, reflecting an increasing recognition on the part of both (creative) writers and (academic) critics that British nature writing currently inhabits a condition of 'feral modernity' that requires the rethinking of conventional aesthetic categories of the beautiful and the sublime.[74] Perhaps the most conspicuous manifestation of this is what might loosely be called the 'Edgelands School' of writers and critics. Directly or indirectly influenced by vitalist understandings of the natural world sometimes bracketed under the equally loose category of 'the new materialism',[75] these writers have

[71] Buell, *Environmental Imagination*, p. 91.

[72] Helen Macdonald, *H Is for Hawk* (London: Jonathan Cape, 2014), p. 265. [73] *Ibid.*, p. 265.

[74] Smith, *New Nature Writing*, n.p.

[75] The new materialism, perhaps better seen as a loose constellation of related approaches than as a unified philosophy or movement, has rapidly evolved into a new orthodoxy in ecocritical thinking, though – like ecocriticism itself – it ranges across a number of different disciplines that may have little or nothing to do with the reading of literary texts. New materialist approaches tend to stress the agential capacity of matter, the co-implication of nature and culture and the entanglement of all (both human and nonhuman) life forms. Most of this book is influenced to some degree by new

tasked themselves with examining those fringe (i.e. semi-urban, semi-rural) areas of the country which, often far from beautiful, tend to be left out of more conventionally embellished loco-descriptive accounts.

The evocative word 'edgelands' is usually credited to the writer-activist Marion Shoard, who uses it to describe 'the interfacial rim' that currently exists between the English countryside and the nation's cities,[76] but derivations of it can also be traced to, for example, Richard Mabey's influential 1973 study *The Unofficial Countryside*, as well as to the work of the various writers and campaigners associated with the late twentieth-century grassroots environmentalist organisation – for which Mabey himself was one of the earliest patrons – Common Ground (see Chapter 3).[77] Following on from this, a number of more recent environmentally oriented writers and critics have tried to make the case for edgelands, less as a hybrid geographical concept than as a ramshackle ensemble of 'new ways of looking and writing [about the environment] that are struggling out of old conventions', and which consequently require a 'self-conscious renegotiation of our [normative] relationship with landscape and place'.[78] These ways of looking and writing are seen as distinctively modern, and indeed they are, but as we have argued, they are probably best viewed in terms of a productive fusion of post-Romantic and late modern sensibilities that allows for the critical exploration of

materialist thinking, though its presence is felt more strongly in some chapters than in others: see especially Chapters 1 and 3. For different perspectives on new materialism(s), see Jane Bennett, *Vibrant Matter: A Political Ecology of Things* (Durham, NC: Duke University Press, 2010) and Diana Coole and Samantha Frost (eds.), *New Materialisms: Ontology, Agency, and Politics* (Durham, NC: Duke University Press, 2010).

76 Marion Shoard, 'Edgelands', in Jennifer Jenkins (ed.), *Remaking the Landscape: The Changing Face of Britain* (London: Profile Books, 2002), p. 117.

77 The arts and environment charity Common Ground was co-founded in 1983 by three members of Friends of the Earth – Sue Clifford, Roger Deakin and Angela King – who shared a concern that the large-scale environmental movements of the 1970s had tended to focus on equally big, publicity-generating issues (wilderness, endangered animals, etc.) at the expense of more ordinary landscapes and species which deserved to be championed as well. Several contemporary critics, notably Joe Moran and Jos Smith, have seen Common Ground as one of the principal inspirations behind the new nature writing, though as we argue in this chapter by no means all contemporary British nature writing is or even aspires to be environmentally activist, and 'new nature writing' is in this as in several other respects a highly misleading umbrella term. Deakin's own creative work (e.g. *Waterlog*, published in 1999) is a case in point, owing as much to the mild eccentricities of mid to late twentieth-century travel writing as to the heightened environmental consciousness of the new nature writing, a label that – despite his faultless environmentalist credentials – seems somehow ill suited to his work. See Roger Deakin, *Waterlog: A Swimmer's Journey through Britain* (London: Vintage, 2000 [1999]); also Richard Mabey, *The Unofficial Countryside* (Dorchester: Little Toller, 2010 [1973]); Moran, 'Cultural History of New Nature Writing'; Smith, *New Nature Writing*. On Mabey's *The Unofficial Countryside*, see also Chapter 3.

78 Smith, *New Nature Writing*, n.p.

multiple coexisting *temporalities* as much as shifting orders and configurations of space. Edgelands, in this last sense, are neither 'new' nor 'old'; rather, they occupy the margins of time – an uncertain temporal hiatus in which the past sits uneasily with the present and in which numerous reflective opportunities are offered to celebrate what is left, but also to mourn what has been lost.

Probably the most popular recent text to limn this territory is Rob Cowen's suitably named *Common Ground* (2015), a partly fictional account of the author's attempts to familiarise himself with the surroundings of his new hometown in post-industrial Yorkshire, surroundings which, as he discovers, may be unsightly on the surface, but which, as Macdonald might put it, eventually turn out to be 'buzzing and glowing with life'.[79] Our more immediate focus in this section, however, will be on a slightly earlier and less well-known text, Paul Farley and Michael Symmons Roberts' *Edgelands: Journeys into England's True Wilderness* (2012), which we have chosen not just for its range of physical and emotional encounters with edgelands, but also for its intellectual engagement with a figure that has become fashionably associated with the new nature writing without necessarily having been put to the test.[80]

Edgelands ostensibly involves a series of playful journeys into what the co-authors call England's 'negative spaces' (p. 76): wastelands and derelict places which, situated at a convenient remove from the cities that have created them, remain largely unlived in and almost totally unloved. These places – landfill sites, gravel pits, abandoned factories – are scanned for what they reveal about the ruinous speed with which we currently live and the tangled mesh of different elements and materials with which we, often destructively, interact. This is 'feral modernity' writ large, its ugliness hiding in plain sight, but as the authors defiantly remark, 'what's so good about wild animals [and] wild flowers [anyway]? All they do is what their instincts tell them, what their genes taught them. Feral means you have a history, a proper backstory. And the edgelands are the domain of the feral' (p. 158).

In this context, *Edgelands* stages a spirited recuperation of edgelands as fascinating repositories of local history, in which what had been previously ignored or discarded is now affectionately restored (p. 210). By visiting these sites and excavating the hidden histories they contain, Farley and Roberts perform a kind of post-industrial rewilding in which the pastoral

[79] Rob Cowen, *Common Ground* (London: Hutchinson, 2015).
[80] Farley and Roberts, *Edgelands*. Further references are in the text.

and the urban, the organic and the mechanical, are co-conscripted for the purpose of bringing a shadowy industrial era, which the authors see as being 'haunted by a future that never happened' (p. 234), uncannily back to life. In this and other respects, *Edgelands* is a quintessentially post-pastoral text, modern to a fault but also punctuated by a recycled Romanticism in which myths and legends of the urban wild are regenerated and the relentless teleology of modern development is held, albeit precariously, at arm's length.

At the same time, the text is an exercise in *literary* recycling, replete with references to both Romantic and Modern artists, and similarly aware of more recent attempts at 'wild writing': at one point Macfarlane's *The Wild Places* is ironically referred to as a nostalgic intertext (p. 118). In this last context, it would be easy enough to stake a claim for *Edgelands* as a postmodern text, and it certainly flirts with postmodernism, though its modish references to Baudrillard and Jameson, in the context of the early twenty-first century, always run the risk of sounding out of date. But as we have previously suggested for contemporary British nature writing at large, it probably makes more sense to see it as late modern, haunted less by a future that never happened than by a past to which it has only limited imaginative access, and threaded through with a series of adolescent memories which, even as they resonate with naïve enthusiasm, cannot help but come across as slightly nerdish – as if the edgelands were being reconstituted as a kind of male playground where responsible grown men can act or, better, can imagine themselves as acting, like naughty boys (p. 148).

In their defence, Farley and Roberts are well aware of their own infantilising tendencies to revert to a romanticised past, just as they play mischievously on Tim Edensor's idea of industrial ruins as latter-day romantic landscapes, 'so atmospheric that [they offer] deracinated human beings a glimpse of the sublime' (p. 153).[81] The overall effect of this is to create a kind of late modern counter-nature writing that intellectually resists the very romantic temptations to which it emotionally subjects itself: the temptation to reminisce about the imagined past, the temptation to champion the wild as a place of neglect and abandonment and, above all, the temptation to search for a 'true' wilderness that defies official designation (the 'protected' heritage area; the 'depopulated' national park).

Similarly, we would argue that there is something bogus about the idea of edgelands as the 'new wild', and about both of these as 'common

[81] See also Tim Edensor, *Industrial Ruins: Space, Aesthetics and Materiality* (Oxford: Berg, 2005).

ground', open to all and free from the clutches of corporate homogenisation (p. 142). Cowen's way of dealing with this is to reclaim the common ground near his house for himself, and to complain bitterly when his possessive fantasies are disrupted. Farley and Roberts' is to be knowing about the limits of their own efforts, as in the self-mocking final sequence in which they expatiate on the abandoned West Pier in Brighton – former home of the entertainment centre 'Laughterland' – only to find that the last laugh is on them: just months after their visit the pier first partly collapses then comprehensively burns down.

Two opposing morals can be drawn about the wild, both linked to Anthropocene thinking. The first is that the wild, a personalised human idea, can sometimes even be a collective human delusion. The second is that the wild, irrespective of what humans want or create for themselves, does what it will. That contemporary British nature writing regularly exhibits both of these traits is an argument for its defence, not a confirmation of its irrelevance. But it is still a reminder that even the most self-reflexive forms of nature writing are likely to have limited environmental impact. Then again, that may not be what they are cut out for. And that's why, contra some of the more pious environmentalist accounts of the new nature writing, some of the most illuminating pieces of nature writing may turn out to be the least useful environmental texts.

Affective Animals

Three animal-based texts by British writers stand out in the twentieth century, all of them predating what we have been loosely describing here as the contemporary period: Henry Williamson's *Tarka the Otter* (1927), Gerald Durrell's *My Family and Other Animals* (1956) and J. A. Baker's *The Peregrine* (1967).[82] (A fourth also potentially qualifies, Kenneth Grahame's ever-popular children's novel *The Wind in the Willows* (1908), but as this is at best a fringe case of nature writing in the way that this book understands it, we are excluding it here.[83]) Two of these texts, *Tarka* and *The Peregrine*, have withstood the test of time and are still strikingly relevant; only Durrell's now comes across as old-fashioned, locked (as argued in the previous chapter) into early conservationist thinking about animals as sources of human entertainment and the privileged eccentricities of the

[82] See Henry Williamson, *Tarka the Otter* (London: G. P. Putnam's Sons, 1927); Gerald Durrell, *My Family and Other Animals* (London: Rupert Hart-Davis, 1956); John Alec Baker, *The Peregrine* (London: Collins, 2015 [1967]).

[83] See Kenneth Grahame, *The Wind in the Willows* (Oxford: Oxford University Press, 2014 [1908]).

Englishman abroad. All three are as interesting for what they conceal as what they reveal, which is unsurprising given that they are the products of deeply troubled human personalities. It has been suggested that the human viciousness of *Tarka* (as well as the brutality of Williamson's real-life politics) can be partly explained by his traumatic wartime experiences in the trenches, while Durrell's semi-fictionalised memoir has been seen as a bright-lit exercise in physical and psychological escapism, responding to personal bereavement by conjuring up the island of Corfu as a kind of anti-England, an enchanted 'looking-glass world'.[84] Meanwhile, Baker's coruscating *The Peregrine*, ostensibly the product of an unremarkable man's obsession with a remarkable creature, asks decidedly uncomfortable questions about the limits of human understanding and intelligence that cannot help but reflect back on the mysterious figure of the author himself.

If this is to make the obvious point that animal writing is necessarily about humans, albeit mostly about human defects and insufficiencies, it is also to suggest that it provides a fascinating record of the human-animal relationship over time. More recently, this changing relationship has been the focal point for a burgeoning research field, animal studies, which has shed new light on both the meaning of the human and the tangled interactions between the human and animal worlds (see also Chapter 3). Much work in this field now focuses on the further, symbiotic relationship between two early twenty-first-century philosophical developments, which – in a questionable kind of intellectual shorthand – are currently referred to as the 'animal' and 'affective' turns. The *animal* turn is based on the premise that animals, rather than functioning as mere projection screens for a variety of human concerns and interests, are self-constituting subjects in their own right and active material agents in their own terms.[85] The *affective* turn, meanwhile, focuses on the various processes by which human and animal bodies (or, perhaps better, human and *other* animal bodies) engage meaningfully with one another in situated ways that are irreducible to set behavioural patterns, and that involve open forms of mutual 'becoming' or interactive 'worlding' that belie the ontological

[84] Durrell, *My Family*, p. 17; see the more extended discussion of Durrell's work in Chapter 3. See also Douglas Botting, *Gerald Durrell: The Authorised Biography* (London: HarperCollins, 2000) and Guy Yeates, *Henry Williamson: The Artist As Fascist* (independently published, 2017).

[85] See, for example, Linda Kalof and Georgina M. Montgomery (eds.), *Making Animal Meaning* (East Lansing: Michigan State University Press, 2012); also Kari Weil, 'A Report on the Animal Turn', *differences*, 21 (2010), 1–23.

stability implied by conventional understandings of the grounded self (see the discussion of Griffiths earlier in this chapter).[86]

These two turns come together in what we will be referring to in this section as the multifunctional figure of the *affective animal*. Affective animals are characterised, not just by their broad emotional appeal, but also by their situated capacity to affect human behaviour. They may also be so named insofar as they are able to elicit public sympathy, and in this respect the figure of the affective animal becomes a kind of emotional cipher, a catalyst for media-generated activist and/or advocacy work.[87] But not all animals function in this way of course, and one of the benefits of animal writing is that it does not restrict itself to so-called charismatic animals that are manipulated in such a way as to tug at our heartstrings or activate the recognition of 'our' complicity in 'their' plight. Indeed, one of the products of what we earlier called the ecological turn (another turn!) in British nature writing has been the renewed attention to less obviously significant species within the larger framework of ecosystems in which – as has been known for several centuries – it is the insects and worms of this world that do the vital work. That these systems are subject to potentially catastrophic breakdown, and are far less stable than they seem, has also been known for quite some time, although today's 'Anthropocene condition' has reinforced this.[88] Hence the elegiac edge in modern nature writing, an edge significantly sharpened by the unfolding consequences of the Anthropocene, which registers both the anticipatory fear of likely disappearance and the felt disappointment of actual loss. And hence another form of regret experienced in texts that reveal, through several layers of mediation, that direct access to animal worlds is increasingly limited, and that – for many species – these worlds may not be accessible at all.

But if the history of British animal writing is fraught in these and other ways, it also fights against its own melancholic impulses. Iconic British animals – badgers and barn owls, ospreys and otters – continue to feature in contemporary texts, often in numbers that belie their actual ecological status; indeed, in some of these texts, for example, John Lewis-Stempel's aforementioned *Meadowland*, long inventories of local species are delivered that play to the best traditions of the Darwinian sublime. Even the more hopeful texts, though, tend to be haunted by the fear of loss, and there are

[86] See, for example, Clough and Halley, *Affective Turn*.

[87] For further reflections on affective animals, see Jopi Nyman and Nora Schuurman (eds.), *Affect, Space and Animals* (London: Routledge, 2015).

[88] Purdy, *After Nature*, n.p.

few indeed that dabble in the kinds of self-satisfied 'bourgeois escapism' for which nature writing is sometimes popularly known.[89] Instead, contemporary British animal writing, combining the long Western traditions of the pastoral elegy and the beast fable, provides a set of more or less anthropomorphic stylistic exercises in human alienation from the natural world. We say 'more or less' insofar as animal writing has historically gestured towards that which it can never achieve – that is, the human attempt to 'become' an animal. Admittedly, some forms of writing have little or no intention of doing this, turning animals into little more than human beings, often vulnerable or put-upon human beings, in numerous entertaining varieties of mock disguise. But other forms, taking their cue from the latest ethological discoveries, offer a more scientific approach to the understanding of animal behaviour, attempting an impossible-but-necessary reciprocity of vision in which human beings are both fashioning observers and sometimes unwilling recipients of a keenly experienced *animal gaze*.

While animal writing, like other forms of nature writing, is by definition multisensory, it is the sense of sight that predominates in most cases, backed by the multifaceted history of empirical observation that has accompanied its most readily recognisable current British form, the first-person natural-historical account. This suggests that the study of animal writing might benefit from the various theories of the gaze that contemporary visual theorists have developed. One important point to bear in mind is that 'the gaze' is not the same thing as 'the look'. For Kaja Silverman, the look describes a visual *category*, often driven by personal desire, whereas the gaze describes a broader visual *regime* that is shaped by dominant social and cultural forces and that frequently operates as an instrument of control and mastery that positions the subject in relation to what it sees.[90] For humans to look at animals, in Silverman's terms, always implies a certain degree of intellectual and emotional projection; within the broader framework of the gaze, it also opens up the possibility of a returned look that confirms the other as a subject in its own right.[91] Within this general context, Wendy Woodward distinguishes between two kinds of animal gaze: an *objectifying* gaze that confirms (imagined) human superiority over animals, and a *reciprocal* gaze that acknowledges

[89] Steven Poole, 'Is Our Love of Nature Writing Bourgeois Escapism?', *Guardian*, 6 July 2013.
[90] Kaja Silverman, *The Threshold of the Visible World* (New York: Routledge, 1996), pp. 134, 156.
[91] *Ibid.*, p. 95.

our 'subjective kinship' with them.[92] Woodward is too committed to the possibility of placing humans and animals in what she calls a 'continuum of kinship';[93] other theorists, notably Donna Haraway, suggest that humans and animals coexist in a much messier entanglement of desires and forces that is by definition unbalanced and asymmetrical, and in which the possibility always exists that human interest in the animal is not reciprocated at all.[94]

What does all this mean in the context of contemporary British animal writing? One thing it means – to return to the work of Purdy – is that animal writing, even given the scientific advantages of the times, remains as much an index of ignorance as a vehicle of insight into the phenomenal lifeworlds of animals; as Purdy starkly puts it, 'we [just] don't know what an animal's life means, to it or to us'.[95] Another thing it means is that this continuing ignorance can be brought to bear on the history of what we have *done* to animals; in this second sense, contemporary animal writing operates within another kind of historical penumbra, that surrounding human guilt. This isn't to turn all animal narratives into moral tales, for as might be expected contemporary texts tend to be critical of their own moralising tendencies, but it *is* to suggest that animal writing, often to a greater extent than its place- or landscape-based counterparts, draws upon a conspicuously uncanny aesthetics driven by simultaneous unease and uncertainty over what it purports to represent.

Finally, a third thing it means is that the various uncertainties and failures embedded within a history of British animal writing are exacerbated by the current conditions of the Anthropocene, which matches the *actual* disappearance of animals – in alarming numbers – to their *metaphorical* disappearance from our everyday lives.[96] In contemporary writing, these two forms of disappearance can easily start to blend in to one another, creating a grey area between *extinction* narratives, which document irrecoverable losses, and *pseudo-extinction* narratives, which build up apocalyptic atmospheres of disappearance even if the science tells us

92 Wendy Woodward, *The Animal Gaze: Animal Subjectivities in Southern African Narratives* (Johannesburg: Wits University Press, 2008), p. 3.
93 *Ibid.*, p. 14.
94 Donna Haraway, *When Species Meet* (Minneapolis: University of Minnesota Press, 2008), p. 24.
95 Purdy, *After Nature*, p. 244.
96 For different versions of this argument, see John Berger 's justly famous essay 'Why Look at Animals?' in his collection *About Looking* (London: Bloomsbury, 2009 [1980]), and Akira Mizuta Lippitt's monograph *Electric Animal: Toward a Rhetoric of Wildlife* (Minneapolis: University of Minnesota Press, 2008).

that this isn't necessarily the case.[97] Recent work on extinction suggests that it is as much a failure of the imagination as anything else;[98] contemporary animal writing fights all the more vigorously against this failure because it is one that it is fearful of repeating itself.

In what follows, we want to chart these various interconnected struggles, which are at once physical and psychological, representational and epistemological, via a sample of older and more recent animal texts. Backing up our earlier reflections on the animal gaze, we will turn first to two pairings of texts that illustrate the potentially visionary quality of modern animal writing while also showing its *limits* of vision in relation to the various 'transspecies encounters' it wishes to enact.[99] In the first subsection, we will compare J. A. Baker's aforementioned *The Peregrine* to Kathleen Jamie's scattered observations on hawks in her 2005 collection *Findings*; we will then go on to make a further comparison between Helen Macdonald's award-winning 2014 memoir *H Is for Hawk* and the book after which it partly styles itself, T. H. White's weird and wonderful *The Goshawk* (1951).[100] In the second subsection, we will turn our attention to a suite of texts that examine the relationship between humans and otters, ranging from two earlier classics, Williamson's *Tarka the Otter* (1927) and Gavin Maxwell's *Ring of Bright Water* (1960), to more recent footsteps journeys (Miriam Darlington's *Otter Country*) and experimental enactments (Charles Foster's *Being a Beast*).[101]

***

All of Russia, it is rumoured to have been said, emerged from Gogol's *Overcoat*. It is worth wondering whether a similar thing might be said of

[97] Thanks to Kate Marshall for bringing the term 'pseudo-extinction' to our attention. Understandings of pseudo-extinction vary, with the most widely accepted, biological definition referring to the extinction of a *parent* species, while an affiliated *sub*species is still alive (see, for example, Rick de Vos, 'Extinction Stories: Performing Absence(s)', in Laurence Simmons and Philip Armstrong (eds.), *Knowing Animals* (Leiden: Brill, 2007), pp. 183–95. Our own understanding of pseudo-extinction, derived in part from Marshall's, differs from this. See Kate Marshall, 'What Are the Novels of the Anthropocene? American Fiction in Geological Time', *American Literary History*, 27 (2015), 523–38.

[98] See Ursula Heise, *Imagining Extinction: The Cultural Meanings of Endangered Species* (Chicago, IL: University of Chicago Press, 2016); also Thom van Dooren, *Flight Ways: Life and Loss at the Edge of Extinction* (New York: Columbia University Press, 2014).

[99] See Donna J. Haraway, *When Species Meet* (Minneapolis: University of Minnesota Press, 2007).

[100] Baker, *The Peregrine*; Kathleen Jamie, *Findings* (London: Sort of Books, 2005); Macdonald, *H Is for Hawk*; T. H. White, *The Goshawk* (London: Weidenfeld and Nicolson, 1964 [1951]).

[101] Williamson, *Tarka the Otter*; Gavin Maxwell, *Ring of Bright Water* (London: Penguin, 2001 [1960]); Miriam Darlington, *Otter Country* (London: Granta Books, 2013); Charles Foster, *Being a Beast* (London: Profile Books, 2016).

contemporary British nature writing and Baker's *Peregrine*, though it is really only recently, with the various high-profile accounts that accompanied the golden jubilee of its publication, that Baker's hallucinatory, sometimes harrowing, rendition of his experiences with the titular animal has come to get the acknowledgement it deserves. Like Gogol's quirky masterpiece, *The Peregrine* is both terrifying and distinctly odd, with both narrative accounts almost willing the real/imagined calamities they visit upon themselves. Famously, *The Peregrine* took ten years to write though the text itself is quite short and only covers six months or so of its author's obsessive tracking of peregrines in his low-key Essex neighbourhood. Robert Macfarlane, one of several contemporary nature writers to live in these birds' shadow, has described the text as written in 'ecstatic, violent, enraptored prose'.[102] It is certainly true that the adjective 'hawk-like' is as apt a depiction of the text's style as it is of its subject, especially in relation to its hyper-concentrated focus on the world-creating, but also potentially world-destroying, properties of the human/animal eye. In fact, *The Peregrine* is a Symbolist work in the Blakean mould, albeit adapted to the late modern (mid twentieth-century) context of a nation still struggling with the collective traumas of two devastating world wars and the hawkish spectre of another, in the shape of a potentially earth-destroying nuclear war.

In this context, to see is alternatively to select one's prey and to serve oneself up as the predator's willing sacrifice. *The Peregrine* moves alarmingly between these two opposing subject positions, never quite becoming the hawk (because its author cannot) but never quite succumbing to it either. At times, Baker seems to claim an almost shamanic affinity with the hawk, as in one close encounter when the hawk's eyes alight on the man's face, and Baker, part unnerved and part mesmerised, consoles himself with the thought that 'he regards me now as part hawk part man, worth flying over to look at from time to time but never wholly to be trusted; a crippled hawk perhaps, unable to fly or kill cleanly, uncertain and sour of temper'.[103] More often, though, Baker admits that he knows little about either the hawk or its intentions. And for all the forensic precision of his observations, he cannot help but elevate the hawk into a kind of deathly symbol. This is all the more shattering in that peregrine falcons at the time

[102] Robert Macfarlane, 'Violent Spring: The Nature Book That Predicted the Future', *Guardian*, 15 April 2017, p. 2.
[103] Baker, *The Peregrine*, p. 146.

were themselves a deeply endangered species: Baker's hawks are relentlessly pursued precisely because he fears they may be England's last.

It would be an understatement to say that *The Peregrine* is fascinated by death; rather it embraces it. Almost lovingly detailed descriptions of fresh kills are intensified in the text by howling imprecations against man the killer:

> No pain, no death, is more terrible to a wild creature than its fear of man. A red-throated diver, sodden and obscene with oil, able to move only its head, will push itself out from the sea-wall with its bill if you reach down to it as it floats like a log in the tide. A poisoned crow, gaping and helplessly floundering in the grass, bright yellow foam bubbling from its throat, will dash itself up again and again on to the descending wall of air, if you try to catch it. A rabbit, inflated and foul with myxomatosis, just a twitching pulse beating in a bladder of bones and fur, will feel the vibration of your footstep and will look for you with bulging, sightless eyes. Then it will drag itself away into the bush, trembling with fear.
>
> We are the killers. We stink of death. We carry it with us. It sticks to us like frost. We cannot tear it away.[104]

Scenes like this are almost too awful to behold, but they fix their raptor gaze on us. We want to look away and cannot, but if we are being honest with ourselves, we want to look as well. By looking, we are trapped into complicity, but like the hawk itself Baker seems indifferent to this. As Macfarlane rightly says, the text 'suggests no basis for the establishment of an environmental ethics born of commonality', though he goes on to cite several instances of contemporary avian conservation initiatives that have been directly influenced by Baker's work.[105] The main reason for Baker's indifference is that he remains firmly locked into a post-Romantic vision of his own solitude. Though we learn next to nothing about him in the text, he is ever-present in it – a helpless witness to his own sensory impoverishment, no doubt, but a witness nonetheless. The struggle for vision in the text (a literal one too: he was myopic) is Baker's own; he is not especially interested in whether we want to participate in it. It is also a struggle that is lost because it was too late in the first place. Whether it is too late or not for mankind is a question the text leaves open for us; but it gives us little reason not to believe that Baker thinks it is too late for himself.

For the Scottish poet and nature writer Kathleen Jamie, *The Peregrine* is, somewhat cautiously, a respected 'classic of English natural history: a keen, affecting book about a man's affiliation and these particular birds'.[106] But

[104] *Ibid.*, p. 113. [105] Macfarlane, 'Violent Spring', pp. 2–3. [106] Jamie, *Findings*, p. 35.

later in a chapter on peregrines and other birds of prey in her 2005 work *Findings*, Jamie is keen to know more: 'Who was this man who could spend ten years following peregrines? Perhaps he was landed gentry. What allowed him to crawl the fields and ditches all day, all winter, until he could tell just by a tension in the air that there was a peregrine in the sky?'[107] As is characteristic for Jamie's work, these questions and speculations go unanswered, though she admiringly suggests that, for a man who would seemingly 'annihilate himself and renounce his fellows', it must have been an 'act of consummate communication to his human kind [to] step back into language and write a book still spoken of forty [now more than fifty] years on'.[108]

Baker's book is perhaps not as generous to its audience as Jamie implies, but she neatly captures its overwrought qualities, the obsessiveness of both the pursuit itself and the language that attends to it, which slavishly imitates the hawkish properties of vitality and hunger that it knows it cannot appropriate for itself.[109] Jamie's own approach to writing is very different: understated, unassuming and always wryly conscious of the gap between the ambivalent freedom accorded to wild animals and the pleasurable constrictions of her domestic life. Above all, Jamie's writing performs the sly trick of managing to say a lot while repeatedly confessing to its own state of ignorance: self-interrogating phrases like 'I have no idea', 'I couldn't say' and 'I'm beginning to wonder' are scattered through the pages of *Findings*, as they are in her other nature texts. (Predictably perhaps, Jamie has publicly doubted whether her work is nature writing at all, or whether indeed 'nature writing' exists as a viable literary category.)[110]

Temperamentally, Jamie's work combines the modern dissociation of the self with an updated form of Romantic negative capability: acting together, these two qualities make for a style of writing that can appear clinically austere but is always charmingly honest and refuses to tie itself down to simple explanations: 'I want to learn to notice, not to analyse', she says at one point in *Findings*, 'to still the part of the brain that's yammering, "My God, what's that? A stork, a crane, an ibis? Don't be silly, it's just a weird heron'.[111] For all Jamie's protestations to the contrary, this is modern nature writing to a fault, as aware of itself as of its subject and,

107 *Ibid.*, p. 43. 108 *Ibid.*, p. 43. 109 *Ibid.*, p. 43.

110 See, for example, her sympathetic review of Tim Dee's *Four Fields* in the *Guardian*, 24 August 2013, in which she begins by calling the concept of 'nature writing' problematic and ends by doubting its existence at all.

111 Jamie, *Findings*, p. 42.

acting as if to follow its own curiosities without feeling the need to rationalise or catalogue them, more likely to find things by default than by design – hence the fragmented structure and simulated randomness, both classic modernist strategies, embedded within her work.

There is an uncanny note to this work as well, as when Jamie – following in Baker's footsteps – describes the peregrines she has recently taken to watching in her Scottish Lowlands neighbourhood as 'flicker[ing] at the edge of one's senses, at the edge of the sky, at the edge of existence itself'.[112] Her vision, unlike Baker's, may not be impaired, but like his, her epistemological horizons are strictly limited: there is only so much that can be known about peregrines, and however attentive we are to their movements, they will always have the capacity to move beyond our field of vision, ascending from approachable objects of scientific inquiry to sublimated subjects of folklore and myth.

Jamie's response to this, however, is to keep her own feet firmly on the ground, and to acknowledge that her world and that of the peregrine may be entangled at one level, but are emphatically not the same. 'I like being able to glance up from my own everyday business', she says, 'to see the osprey or the peregrine go about hers'.[113] This frisson of recognition is then intensified by her knowledge that these sightings are relatively rare occurrences. 'What it is about the peregrines is their rarity. I'm sure they're gone now, but for a while I enjoyed the pleasure of a warm secret: I could watch the uncommon and handsome bird from my own window, and know it was there.'[114] That these 'warm secrets' are shared with us is part of the appeal of Jamie's writing, which ultimately turns its back upon the 'falconer's worship of mastery, [and] the wish to identify with terror, with the predator over its prey'.[115]

A rather different examination of the masculinist myths surrounding falconry can be found in the work of T. H. White, still best known today for his suitably garish Arthurian romantic novels, but whose scarcely less melodramatic nature memoir *The Goshawk*, written in the immediate Second World War period, has come to acquire a minor cult status of its own. *The Goshawk*, the story of its solitary author's increasingly manic attempts to train the titular bird, is probably most interesting today for the glaring contradictions that accompany a hysterical anti-modern crusade written almost entirely from a modern sensibility, and explosively bringing together Romantic and Modern imaginaries, the one fuelled by ancestral longing, the other by present misery and despair. The hawk itself is an

[112] *Ibid.*, p. 47. [113] *Ibid.*, p. 46. [114] *Ibid.*, pp. 46–7. [115] *Ibid.*, p. 46.

ancestral figure, the living embodiment of its excitable author's 'excursion [out] into the fields and back into the past'.[116] This untutored excursion, and the equally amateurish training it involves, are likened to a mock trial in which White doubles as unwilling executioner and tormented victim, exchanging these roles as often as the 'trained' hawk turns from flesh-shredding apex predator to human symbolic prey.

As the tale proceeds, this trial turns into a kind of medieval mortification rite, a submission of the self to punishment for the sake of others: one which, in a not entirely dissimilar manner to Baker's *The Peregrine*, shoots off contemporaneous sparks of modern warfare and civilisational collapse.[117] Also like Baker, White clearly envies the hawk – for its hugely powerful presence, for its vastly superior powers of vision – but for these self-same reasons he is terrified of it, acutely aware that 'hawk vision' allows for the imagination of killing, but also of being killed.[118] As these fears increase, the hawk becomes a powerful vehicle for White's frustration and pent-up rage: against 'civilised' society, against the warmongering modern world, but most of all against himself. 'What right had a cowardly recluse who fled from his fellow men . . . to write about these almost fabulous creatures?'[119] Here as elsewhere in the text, White's feeble attempts at hawk training turn into a post-Romantic exercise in self-estrangement: for it is not the hawk that is other, but White (who likens himself at one point to Victor Frankenstein) who is other to himself.

There's not much to be learned about hawks in the text, and White freely admits that he doesn't know much about them either; rather, the text enacts a semi-parodic mating of raging souls in which the hawk, 'haunted by moods and mania', becomes an objective correlative for White himself.[120] Indeed, *The Goshawk* is stalked by madness from beginning to end, a condition that goes far beyond the foolishness of its own enterprise, gesturing instead towards a collapsed world order that can neither be redeemed nor rectified, and over which the image of the hawk magisterially presides as a latter-day Attila figure, a legendary tyrant who 'immolated victims, sacked cities, [and] put virgins and children to the sword'.[121]

White's livid martial fantasies are a far cry from the more intimate scenes of transspecies encounter we find in Helen Macdonald's 2014 memoir *H Is for Hawk*; nonetheless, *The Goshawk* is a formative influence on Macdonald, whose text deliberately situates itself in White's shadow and

116 White, *The Goshawk*, p. 16. 117 *Ibid.*, p. 52. 118 *Ibid.*, pp. 67–8. 119 *Ibid.*, p. 75.
120 *Ibid.*, p. 99. 121 *Ibid.*, p. 172.

consciously engages with the estranged modern condition that it both dolefully inhabits and energetically attacks. Reversing the temporal order of their respective texts, Macdonald refers to White at one point as a projected Merlin figure, 'an imagined future self'.[122] It's clear though that White, the story of whose mental and physical turmoil is interwoven with Macdonald's memoir, is less the legendary magician of his own romantic imagination than a deliberately unbalanced modern figure, the unstable origin of a calculated negotiation between an unwanted (war-torn) future and an equally violent past.

In this context, *H Is for Hawk* is much more than the female-centred recovery narrative it has sometimes been taken to be; rather, it is a failed attempt at human–animal reconciliation mapped onto an untimely contemplation of the fragmented modern self. This isn't to say that the text doesn't operate at one level as a chronicle of bereavement, but it is melancholic rather than mournful, and the moments of peace and communion it finds are essentially Woolfian 'moments of being', temporary discoveries that reconfirm an overarching sense of loss.[123] As in *The Goshawk*, the struggle for self-control remains unresolved, and the text fights in any case against its own 'strategies of containment' (Jameson).[124] It wants the wild, which is symbolised by the hawk, but knows full well it can't have it; continuing in the search to have what it can't have, it deliberately places itself in harm's way, fearing that the opposite of 'wild' isn't 'tame', but rather that deadening aspiration of modern bourgeois civilisation, 'safe'.[125]

In this sense, *H Is for Hawk* is what the late German sociologist Ulrich Beck might have called a 'risk narrative', in which the authenticity of modern lived experience is gauged by the capacity to exhibit one's own vulnerability to external forces: as Beck apodictically puts it, 'I risk, therefore I am.'[126] What is primarily at risk though, as in White's prior text, is not physical but mental well-being. And, as Macdonald comes increasingly to recognise, the wildness for which the hawk has been made

[122] Macdonald, *H Is for Hawk*, p. 247.
[123] *Ibid.*, pp. 106, 117, 181. For a discussion of Woolf's work, see Chapter 3.
[124] 'Strategies of containment', as Fredric Jameson explains in his work, are ways of dealing with contradiction by constructing substitute truths that limit the damage history does to us: see, for example, Jameson, *The Political Unconscious: Narrative As a Socially Symbolic Act* (New York: Routledge, 2002). We are using the term loosely here to describe White's attempts to control contradiction by accommodating it – strategies that are inevitably, and indeed conspicuously, unsuccessful in his work.
[125] Macdonald, *H Is for Hawk*, pp. 143, 189.
[126] Ulrich Beck, *World at Risk* (Cambridge: Polity Books, 2009), p. 5.

to stand is less a panacea than an illusion; it is also a fundamental misunderstanding of how the hawk lives – not that this understanding can ever be fully arrived at any more than the category of the wild itself is available to human cognition.[127]

What this leaves us with, we would suggest, is a necropolitical rather than biopolitical approach to the human–animal relationship that is played out through *initiation*, a secret if publicly shared rite bound up with death and sacrifice, in which to encounter the wild (animal) is also to conduct a 'conversation with death'.[128] All four texts we have just been examining here are initiations of a kind, although it is arguably only in the earlier texts, White's and Baker's, that the animistic dimensions of their human–animal encounters are fully acknowledged and played out. These initiations, though deeply personal, elevate the texts to the status of myth. They also convey an element of obsession or addiction that is the flip side of a modern 'dissociated sensibility' (Eliot) which, resituating the self as other, causes us to seek, with seemingly increasing desperation, reanimating connections with others' lives.[129] Modern animal writing registers precisely this attempt, not so much to recapture what has been lost, but to find an animating drive, discovered in our encounters with (other)

[127] Macdonald, *H Is for Hawk*, p. 275.

[128] *Ibid.*, p. 209. 'Necropolitics' is the Cameroonian philosopher Achille Mbembe's term for what he calls 'the subjugation of life to the power of the death', a subjugation he sees as exceeding the regulatory bounds of biopolitics (Mbembe, 'Necropolitics', *Public Culture*, 15 (2003), p.39). We are using the term here to convey the sense in which animal writing is bound up in practices of death and sacrifice that often have a quasi-ritual aspect. For an in-depth treatment of this, see Cary Wolfe, *Animal Rites: American Culture, the Discourse of Species, and Posthumanist Theory* (Chicago, IL: University of Chicago Press, 2003); also Nicole Shukin, *Animal Capital: Rendering Life in Biopolitical Times* (Minneapolis: University of Minnesota Press, 2009). In Helen Macdonald's more prosaic but no less terrifying terms, human encounters with animals are 'conversations with death' and animal writing, operating as one of the primary vehicles for such encounters, often examines the 'necropolitical' relations behind them (*H Is for Hawk*, p. 209). New materialist thought has generated a fresh interest in animism as part of the wider transformative process by which bodily exchanges (both human and nonhuman) enact material and spiritual transfers from one state to another, including the ritual passage between death and life. There is no scope to examine this further here although it would be fruitful to do so, and both animal writing and nature writing more broadly stand to benefit from the necro/biopolitical approaches adopted in Wolfe and Shukin's (and, to a lesser extent, Alaimo and Haraway's) work.

[129] The term 'dissociated sensibility', now used in contexts that take it far beyond its original usage, derives from T. S. Eliot's 1921 essay 'The Metaphysical Poets', where he uses it to describe a newfound sense of the separation of thought and felt experience in some seventeenth-century literary works. The splitting of the self that ensues has since been seen as a hallmark of the modern (Western) sensibility, and whether this is attributable or not to Eliot, it certainly applies to modern British nature writing, in which – as we are arguing in this chapter – the 'dissociated sensibility' is writ large.

animals, that might give temporary meaning and energising significance to our everyday lives.

This drive is as likely to be oriented towards death as towards life: hence the double coding of a great deal of contemporary British animal writing, which seeks affective solidarity with animals that kill for survival rather than sport, and whose death instincts operate as risk-filled registers for revitalised human lives. This uncompromising view comes at a price, though it might be going too far to accuse such writing of nihilism; rather, in cases such as Baker's at least, it conveys an immense frustration with human foolishness and arrogance that is reflected at a visceral level in the violent human-animal (and animal-animal) encounters that are retrospectively rehearsed in the text. In the next subsection, we reflect further on the violence encoded in recent and contemporary British animal narratives, associating it this time with otters, not the most obviously violent of creatures, but ones that feature in some of British nature writing's most conspicuously violent texts.

***

The obvious place to start is with Henry Williamson's *Tarka the Otter*, first published in 1927 but still as fresh and unsettling as ever, its instilled violence intensified by recent biographical accounts of its traumatised author's life.[130] Read today, *Tarka* comes across as an X-rated beast fable in which otters, often as violent as the human hunters who see them as pests, feature as hyperkinetic creatures, always on the move, constantly alert, driven by hunger but also by what appears – to human eyes at least – like a permanent condition of jittery nerves. *Tarka* is a novel, though hardly one made up: the product of thousands of hours of painstaking field research by Williamson, it is as much an unsparing portrait of Dartmoor, a once-beautiful landscape increasingly scarred by human use, as a searing indictment of human injustices towards animals (and, though this is not a central feature of the text, towards each other and themselves).

Tarka is also very much a crisis text: describing the fateful collision of 'ancient' and 'modern' worlds, it both simulates an ancientness it cannot grasp and inhabits a modernity it cannot tolerate, with its headlining otters, unable to find a place in either, the very embodiment of what Zygmunt Bauman, in an updated urban context, has described as 'modernity's outcasts', condemned to eke out 'wasted lives'.[131] If this

[130] See, for example, Guy Yeates ' starkly titled *Henry Williamson: The Author As Fascist*, independently published in 2017.

[131] Zygmunt Bauman, *Wasted Lives: Modernity and Its Outcasts* (Cambridge: Polity Press, 2003).

makes the text seem sentimental, it is anything but, and violence is its standard commodity: one typical sequence witnesses the charmingly named Izzywig the hedgehog ('vuz-peg' in local dialect) unceremoniously turned over, protective spines demobilised, and devoured from the inside.[132] To be sure, parts of the text, registered from the otter perspective, read like a celebratory local ethnography of otter culture, replete with Devonian inflections and Kiplingesque gestures towards zoological primitivism (which, as in *The Jungle Book*, is primarily intended to confer dignity on the animal world). But the violence is unrelenting, and in Tarka's post-primitivist world it really does amount to eating or being eaten; there is little time to ponder alternative courses of action, and indeed little time to do anything at all, such is the constant race to find the psychological wherewithal as well as the physical opportunity to survive.

Published more than thirty years after *Tarka*, Gavin Maxwell's *Ring of Bright Water* is similarly written out of an intensely disappointed (in Williamson's case, quasi-misanthropic) apprehension of the violence inherent in modern 'civilised' life. In Maxwell's case, this leads him to Camusfeàrna, a lonely made-up place on the north-west Scotland coast whose real location, which would later ironically become a major tourist draw, is withheld from the reader lest its 'identification in print [should] bring nearer its enemies of industry and urban life'.[133] This is the stage, Maxwell tells us in the foreword, for an escape from 'the prison of adult life' in which the imaginative possibility is explored of looking at 'some portion of the earth as it was before [man] tampered with it', and in so doing of restoring the broken connection modernity has created between mankind and 'the other living creatures of the world'.[134]

Some readers (and *Ring of Bright Water* remains one of the best-selling nature books of all time) may take Maxwell at his word, but many, we suspect, are wary of him; indeed, it is the stereotypically idyllic quality of Maxwell's scene-setting which – and not just for contemporary readers who know the story already – creates an initial impression of unease which quickly gathers momentum in the text. The two otters Maxwell brings into his household (first Mijbil, then Edal) are the temporary stars of the show, but just as they are never entirely trainable, the trust that builds up between human and animal exists under a permanent shadow that belies the brightness and light-heartedness that are surface features of the text. To put it bluntly, the Camusfeàrna project is doomed from the start, not just because there is no more chance of understanding the otters than of

[132] Williamson, *Tarka the Otter*, p. 44. [133] Maxwell, *Ring of Bright Water*, p. 5. [134] *Ibid.*, p. 5.

maintaining the enchanted status of their coastal sanctuary, but also for the obvious reason that otters, however charming they look and entertainingly they behave, are not, nor were ever meant to be, household pets.

Seen in this context, the greatest violence in the text is not the bludgeoning of Mijbil, heart-rending though it is, but the decision to buy and keep him in the first place; in this and other respects, *Ring of Bright Water* joins other human-animal encounter narratives of its period, for example, Durrell's aforementioned *My Family and Other Animals*, as a kind of anti-model for modern animal conservation – not that it was ever intended to be a conservationist text. Whatever the case, *Ring of Bright Water*, read today, lacks the altruistic charm it seems to have had for contemporaneous readers; instead, it is more likely to come across as a damaging portrayal of the possessive individualism of Maxwell himself. There is a kind of childish petulance at work in the text, such as when Maxwell, proud of having achieved the childhood goal of owning an animal that bears his own name (a 'Maxwell's otter'), curses any 'odious taxonomist of the future' who might dare take that name away from him.[135] In his defence, there is gentle self-mockery too, such as in one later episode when, out on one of his morning walks with Edal, he whistles to bring her back only to find himself confronted instead by a wild otter, 'staring at me with interest and surprise'.[136] Seeing that he is no threat, the otter sticks around for another look, then 'resume[s] his leisurely progress southward along the edge of the rocks'.[137] Edal, meanwhile, is left to forage for crabs, small fish and shrimps; when she gets back 'home', she will probably be rewarded with some eels, served up alive to her in her makeshift training tank.[138]

The passage offers a more affecting account of the consequences of captivity than any of the partly staged calamities, which accumulate quickly later in the trilogy, that befall human and animal victims alike. It is important not to be sanctimonious. *Ring of Bright Water* is a product of its time, and – as we have seen in this section – it is probably more revealing of human frailty than of animal vulnerability, though both of these are tangled up with one another as well as with a number of material practices and agencies (local fishing and agriculture, the national tourism industry, the global wildlife trade) that turn it from the affecting story of one man's encounter with wild animals into a modern multi-participatory text.[139]

[135] *Ibid.*, p. 103. [136] *Ibid.*, p. 153. [137] *Ibid.*, p. 153. [138] *Ibid.*, p. 154.

[139] See Haraway, *When Species Meet*, for further reflections on 'multispecies' writing of this kind.

It still seems worth wondering, though, what kind of knowledge is transmitted by books such as Maxwell's (and, in a rather different context, Williamson's) which, for several generations of child and adult readers, have helped to shape an understanding of British otters' still tenuous place in the modern world. As Miriam Darlington, to whose *Otter Country* (2012) we now turn, states, 'much of what we think about otters [today] is mediated by television and by books like *Tarka* and . . . *Ring of Bright Water*'.[140] In the 1960s, she continues, 'keeping pet otters made Maxwell famous. His lyrical account of life with otters in the wild emptiness of the Highlands was highly seductive, but how much did it actually tell us about otters? The enraptured descriptions reveal nothing of the conflicts that emptied the land of its people and little of the reality of keeping wild animals in captivity.'[141]

This is true to a large extent of *Ring of Bright Water* if not of *Tarka*, which as argued earlier is in line with a great deal of twentieth-century British animal writing precisely for what it reveals about the violent modern world in which humans and other animals interact. Darlington's search for the animal behind the tale is thus possibly based on false premises, though this is partly the result of her chosen medium, the footsteps narrative, which is by definition belated with respect to the real and imaginative journeys it seeks to re-enact. Footsteps narratives are caught in a cleft stick between critique and commemoration, with involuntary reverence one potential end result. They are also good examples of what we previously called 'post-memory texts', that is, texts that rely on representation to make up for their lack of direct experience. This doesn't mean that works such as *Otter Country* are condemned to one form or another of experiential inauthenticity, but rather that they depend, to an even greater degree than the works they follow, on an understanding of what the visual theorist Steve Baker calls the 'symbolic availability' of animals as instruments for human needs and desires.[142] As Baker suggests, 'Attitudes to living animals [today] are in large part the result of the symbolic uses to which the concept of the animal is put in popular culture. This is not to set the representational or the popular-symbolic "over" the real. It is simply to say that any understanding of the animal is inseparable from knowledge and its cultural representation.'[143]

[140] Darlington, *Otter Country*, p. 18. [141] *Ibid.*, p. 18.

[142] Steve Baker, *Picturing the Beast: Animals, Identity, Representation* (Champaign: University of Illinois Press, 2001), p. 5.

[143] *Ibid.*, p. 25.

Otter Country, seen in this context, is not one but a variety of cross-hatched quest narratives, in which different temporalities coincide, real and imagined animals co-participate, and the larger symbolic figures of 'the otter' and 'the animal' itself are productively engaged with as ways of both decentring and paradoxically *re*centring the human subjects of these narratives, creating a multispecies mixture of biography and autobiography at the core of which stands the questing author herself.[144] This nesting of narratives probably makes the text sound more experimental than it is, and much of the book is a fairly standard account of Darlington's energetic search for otters across some of the more remote places of Britain, seen alongside and in conversation with her analysis of Maxwell and Williamson's writing, which is in turn enmeshed with their everyday working lives.

Human and animal subjects prove equally elusive, however, and Darlington stresses the difficulties of a multi-scalar tracking process that involves the imaginative crossing of time as well as the material crossing of space. Failure would be too strong a word to describe the result, but like other contemporary writers Darlington is well aware that there is only fleeting access to the creature 'in the flesh', and that most of the time the human quest for the animal turns full circle. This dilemma is writ large in what is probably the greatest temptation of all in animal writing – performatively to 'become' the animal that one can only imaginatively represent. A late scene in *Otter Country* demonstrates this nicely. Near her home in Devon, Darlington discovers a set of otter tracks, and briefly

[144] The idea that we live in a 'multispecies' world has become a critical orthodoxy well beyond the more specific field of animal studies, with several recent interventions in anthropology, biology, geography, literary criticism and philosophy, disciplines that are already incorporated to some extent into the methodological protocols of animal studies itself. The term 'multispecies' (usually unhyphenated) refers to a shared method as much as to a shared world, in which theorists and practitioners attempt systematically to break down what Val Plumwood calls the 'hyperseparations' that have historically regulated human-animal relations, reinforcing the boundary between them and consolidating ideas and illusions of human mastery over the natural world. Multispecies assemblages of different kinds have also been integral to new materialist thought, at least some of which is closely affiliated with posthumanism, itself a collage of related methods that is explicitly designed, not just to challenge the idea of human centrality, but to question the idea of the human itself. One might expect nature writing to be inherently multispecies, but this is in fact far from the case, even in its more theoretically sophisticated strands. That nature writing is human-centred through and through is one of the premises of this book, though there is evidence, by no means restricted to the new nature writing, of a continuing conflict between 'egocentric' and 'ecocentric' sensibilities that helps to define its critical edge (Axel Goodbody, 'From Egocentrism to Ecocentrism: Nature and Morality in German Writing in the 1980s', in Catrin Gersdorf and Sylvia Mayer (eds.), *Nature in Literary and Cultural Studies: Transnational Conversations on Ecocriticism* (Amsterdam: Brill/Rodopi, 2006), pp. 393–415).

imagines what it must be like to be an otter. Crouching in the mud, she tries to recreate the view an otter might have as it comes out of the marsh.

> At the edges [of the river], water and mud lose themselves together. Higher up, oak trees and rock are moored to the bank, and there are dry places between trunks to curl up. I place myself in the crook of a rock and a woody root, and my nostrils fur with earth-odours. If I were a hungry otter I would follow the rim of the crumbling wall and bank, blending myself with the contours, and later drop down a level to the rocks to search for crayfish and eels as they emerge on the rising tide.[145]

Sensing that she is close to the otters, she goes in after them, swimming otter-style with her nose at water level.[146] The otters themselves never appear, and the scant reward for her efforts is to catch two human swimmers, youths skinny-dipping like her in the murky river, embarrassingly unawares. This unsolicited return to the human world makes perfect sense as an ending to the narrative; it also reinforces the delusion of 'becoming an animal' the text intermittently presents. One last text in this subsection not only explores this delusion, but also actively performs it. This is Charles Foster's astonishing 2016 thought experiment, *Being a Beast*.

Being a Beast is nothing if not a risky book. The author's note at the beginning says it is motivated by the dual desire 'to know what it is like to be a wild thing'[147] and to counteract the mock-triumphal tendencies of nature writing, which Foster sees as 'generally [being] about humans striding colonially around, describing what they see from six feet above the ground, or about humans pretending that animals wear clothes'.[148] He posits an alternative: to 'see the world from the height of naked Welsh badgers, London foxes, Exmoor otters, Oxford swifts and Scottish and West Country red deer', but in so doing to chart the olfactory and auditory rather than the purely visual dimensions of animal life.[149] The problem with this approach, as Foster well knows, is that it is impossible. *Being a Beast* is the confessional record of that impossibility, played out across a series of madcap scenes (Foster 'as' badger, 'as' fox, 'as' otter, etc.) that have been read by some reviewers as funny but are best seen as an awkward cross between the plaintively melancholic and the patently absurd.

The absurdity of the exercise doesn't mean it's meaningless. Foster is genuinely interested in how animals smell, touch and feel, and how they negotiate the world from their own species-specific vantage points. He is

[145] Darlington, *Otter Country*, p. 333. [146] *Ibid.*, p. 335. [147] Foster, *Being a Beast*, p. xi.
[148] *Ibid.*, p. xi. [149] *Ibid.*, p. xi.

also genuinely committed to trying to live the lives they lead, whether it is living in a hole and eating slugs (like country badgers) or wandering the streets at night and foraging in rubbish bins (like city foxes). Moreover, as a trained vet he is knowledgeable about animal cognition and physiology, and in this sense *Being a Beast* is empiricism with a difference, backed by detailed scientific observation if still primarily devoted to experimenting 'first-hand' with the (animal) senses so as to develop a heightened (human) sensory appreciation of the phenomenal lifeworld.

The imagination also obviously comes into play here, and indeed Foster describes his book as a 'sort of literary shamanism'.[150] It is shamanism of a very different order, however, than what we arguably see in the work of, say, Baker and Williamson, or in Griffiths' fired-up attempts to tap into the primordial energy she sees as embedded in the psychophysical condition of the wild. Shamanism implies the visionary capacity to transform and/or be transformed – to cross the boundaries between different worlds – but Foster knows he is trapped inside his human body. He can no more become the animals whose actions and behaviour he attempts to copy than he can unlock the secret behind 'becoming-animal', Deleuze's dizzying post-ontological vision – massively influential for many contemporary ecologically minded critics and philosophers – of the incessant 'becomings that compose different [human/animal] bodies'. This vision effectively turns *all* creatures, including ourselves, into affective animals, part of a larger network of ecological relations that continually constitutes and reconstitutes itself.[151]

Not that Foster doesn't try: there are moments in the text, such as when, after spending a summer living water-bound 'as' an otter, he manages to convince himself that he is a shaman after all.[152] But the approaching winter brings him back to earth, and his creeping sense of futility returns:

> This tells me something worth knowing. A depressed shaman, hunter or naturalist can't work at all. A grey soul, apparently, can't penetrate that thin veil between the species. I don't understand the metaphysics. But it seems that you have to be sufficiently 'I' to be another, and depression erodes the 'I' below the critical point. Perhaps, for a human, being an animal is just an extreme mode of empathy – no different in kind from what you need to be

[150] *Ibid.*, p. xi.

[151] Brett Buchanan, *Onto-ethologies: Uexküll, Heidegger, Merleau-Ponty, and Deleuze* (Albany, State University of New York Press, 2008), pp. 159, 161. For the original discussion of 'becoming-animal', see Gilles Deleuze and Félix Guattari, *A Thousand Plateaus: Capitalism and Schizophrenia* (Minneapolis: University of Minnesota Press, 1987).

[152] Foster, *Being a Beast*, p. 93.

> a decent lover or father or colleague. When you're depressed you might simply be nursing that injured 'I' too obsessively to have the energy or attention necessary for empathy. Our nursing strategies are radically misconceived. They all tend to be based on the disastrous misconception that if you give away your Self there will be less of you. In fact of course (as we know when the sun shines), the very opposite is true.[153]

Confessional moments like this give the book an elegiac feel, as if it is already apologising in advance for the diminished empathy and stunted sensory perception its melancholic author sees as the inevitable by-products of contemporary half-living in a 'post-natural' Anthropocene world. His main counter-strategy seems to be a visceral form of self-abasement, but this – by his own admission – is only temporary, and the text is often marked by a glaring (if also guilty) *lack* of empathy, seen in his 'Otter' chapter in his high-handed treatment of both Williamson and Maxwell, and in his forked-tongue revelation of the one advantage he has 'over those true masters of otter writing: I don't like otters very much'.[154]

A few taut pages on otter physiology follow, in which otters are uncompromisingly described as pound-for-pound killing machines, 'jangling, snarling, roaming, twitching bundles of ADHD'.[155] By the end of the chapter Foster has calmed down, his point already having been made, but his mood will swing again in the next chapter, and the next. In this respect, *Being a Beast* is less a thought experiment than a mood experiment in which different animal species are mapped, over and against the reassuring conventions of the 'nature cure' (Mabey), onto its human author's continuously fluctuating neurological chart.[156] Odd though it is, Foster's writing is in fact a logical progression of the self-lacerating kinds of modern animal writing we have been looking at in this section, which often tend to admit the impossibility of empathetic relations with (other) animals *even if this is precisely what they seek.* To some extent, as we have been trying to explain, this is an effect of the soul-destroying modern conditions with which this writing grapples rather than a consequence of the repeated failure to cross the species barrier; more specifically, it is connected to the modern dissociated self.

The ecological orientation of much of this writing – and of contemporary nature writing as a whole – is not an antidote to these conditions but

[153] *Ibid.*, p. 93. [154] *Ibid.*, p. 71. [155] *Ibid.*, p. 73.

[156] See Richard Mabey, *Nature Cure: A Story of Depression and Healing* (London: Vintage, 2008 [2005]). For more on the 'nature cure' genre and its implications, see Chapter 1.

a symptom of them. As has been argued throughout this book, the lyrical appreciation of place was only ever a small part of British nature writing; in its animal-centred counterpart, the same can be said of the locally situated – if globally inflected – appreciation of animals themselves. Instead, what drives both is a general apprehension of loss that carries over in more recent work into an uncanny feeling of lateness: the idea, as previously mentioned, that the time is either out of joint or may be running out. (In some apocalyptic strands, time may well be felt to have run out already.)

If this is an *ecological* idea, translated into the contemporary mantras of the new materialism and the Anthropocene, it is also at heart a deeply *human* one. This means modern animal/nature writers are condemned to repeat what Foster calls the 'two sins [that] have beset traditional nature writing: anthropocentrism and anthropomorphism'.[157] A third sin might be added, closely connected to both: the *pathetic fallacy*. The third and final section of this chapter will look briefly at how contemporary British nature writers have responded to the temptation to project their moods and feelings onto an 'external' world the separate status of which is now increasingly challenged. In some cases, they have done so by addressing that most British of preoccupations, the weather. What 'weather-worlds' (Ingold) do these writers imaginatively portray as well as physically engage with?[158] If these worlds are all tangled up in one another, what future is there for the pathetic fallacy, which is an index of their separation? And what are we to make, in an age of devastating climate change, of the everyday idea of 'weather' itself?

Weather-Worlds

The English weather is more than a national pastime; it is a repository of shared values and experiences, an index of the changing national consciousness over time.[159]

This consciousness is no more stable than the weather is; on the contrary, both are reliably unreliable. As Alexandra Harris demonstrates

[157] Foster, *Being a Beast*, p. xii.

[158] For Ingold's own definition of 'weather-world', see the next section of this chapter. For a lucid discussion of 'weather-worlds' and other related terminology, see also Mike Hulme, *Weathered: Cultures of Climate* (London: Sage, 2016).

[159] A signal irritant of literary criticism in the United Kingdom, discussed intermittently throughout this book, is its abiding tendency to use 'British' and 'English' interchangeably – a tendency that current 'archipelagic' approaches to nature writing are vigorously contesting but have yet fully to counteract. One possible explanation for this is that 'Englishness' continues to carry reassuring connotations, quite often played up in nature books, designed to appeal to a conservative-minded readership; another is that distinctions between 'English' and 'British' – much to the annoyance of

in her magisterial literary-historical study *Weatherland* (2015), there is a strong link between the immateriality of the weather – between phenomena we can see but cannot touch – and our thoughts and feelings: 'the most immaterial parts of ourselves'.[160] These thoughts and feelings are both individual and collective, which is why weather is at once a local, a regional and a national phenomenon. And it's also why weather has proven to be as much an idea (as is the nation itself) of what binds us together as a people, as an experiential register of what happens to us as individuals in any given place at any given time.

If weather is reassuring – or at least reassuringly unsure – the same cannot be said of climate in an era of global warming, which is now generally if by no means uniformly seen as the greatest planetary crisis of our times.[161] As Eileen Crist observes, despite increasingly desperate protestations to the contrary there is 'no longer even a semblance of debate about the reality of global warming, its cause, and the climate change it [both] effects and portends'.[162] The debate has shifted instead to the different possible *consequences* of global warming, with the arcane mechanisms of climate prediction having generated a multitude of different possible scenarios that fluctuate, with the seeming randomness of climate itself, between 'the controllable and the catastrophic'.[163] Climate change has also engendered a variety of cultural responses, some of which shatter

Scottish, Welsh and (in a British Isles context) Irish writers and readers – are quite often blurred in the academic as well as popular discourses applied to these works. A third refers to weather more specifically, which has long been used to nourish national myths, and which feeds in turn into national 'temperaments' and 'sensibilities'. Finally, a fourth indicates the conflicting sentiments that surround nature writing as a heritage form, used whether explicitly or implicitly for the purpose of maintaining existing social hierarchies. Whether the writing *itself* does this is moot; our own speculative view is that its affiliations are mostly conflicted, although these conflicts (and the affiliations behind them) have changed considerably over time. Less exceptionable, perhaps, is that nature writing in the United Kingdom has supplied the ideological glue that holds different communities together, though even here its adhesive qualities are far from guaranteed and – as argued throughout this book – it is often dissent rather than consensus that is at the heart of these works. For more on the ideological tensions that continue to be inscribed within British nature writing, see the Afterword to this book. For further discussion of the British/English nexus, see also Chapter 3.

160 Alexandra Harris, *Weatherland: Writers & Artists under English Skies* (London: Thames and Hudson, 2015), p. 11.

161 For a representative (albeit quirkily expressed) position on this, see Timothy Morton, *Hyperobjects: Philosophy and Ecology after the End of the World* (Minneapolis: University of Minnesota Press, 2013).

162 Eileen Crist, 'Beyond the Climate Crisis: A Critique of Climate Change Discourse', *Telos*, 141 (2007), 29–55 (p. 29).

163 *Ibid.*, p. 30.

conventional temporal and spatial as well as aesthetic categories: the past, the present and the future; the local, the national and the global; the picturesque, the beautiful and the sublime.[164]

Given the epochal importance of climate change, and the environmentalist aspirations of much of what passes for the new nature writing, one might expect to find lengthy descriptions and/or discussions of global warming across a suitably wide range of contemporary literary works. But this is not the case, with one or two notable exceptions. There are at least two good reasons for this. The first reason is that global warming, along with climate change more generally, is difficult to express in aesthetic (especially visual) terms, though recent arguments that standard literary forms such as the novel are fundamentally unsuited to portraying climate change are overstated, while nature writing has a long if not always scientifically literate history of engaging with climate change.

The second reason, pushing slightly against the first, is that *weather* rather than *climate* has tended to be nature writing's most congenial conversation partner. Insofar as one of its main goals is to train the senses towards what Allen Carlson calls 'an aesthetic appreciation of natural environments', nature writing in Britain and elsewhere has been instrumental in showing generations of readers, not only how they are affected by the weather, but also how they themselves help to shape it: through their imaginative perceptions as well as their material recordings of it; and through the many creative ways in which weather, accumulating over time into an intricate tracery of changing moods and memories, is transformed into a vast, emotionally enriching archive of collective national lore.[165]

Whether this archive will stand the test of time remains to be seen. Harris' poignant conclusion to her book is that we may yet need to brace ourselves for the 'last years of English weather' as many of us have known it and to seek consolation instead in the various memories we carry of it, 'the [various] places we have been to or imagined' in relation to it, and 'the [various] books we have read [about it]' during the course of our lives.[166] The archive will likely hold, then, but precisely because of its nostalgia value, just as the 'English weather', its fickleness vouchsafed for centuries, is

[164] For a detailed discussion of this scrambling of temporal and spatial coordinates, see Timothy Clark, *Ecocriticism on the Edge: The Anthropocene As a Threshold Concept* (London: Bloomsbury, 2015).

[165] Allen Carson, *Nature and Landscape: An Introduction to Environmental Aesthetics* (New York: Columbia University Press, 2008), p. 1; see also Harris, *Weatherland*, p. 13.

[166] Harris, *Weatherland*, p. 389.

already well on its way to being an officially commemorated – and therefore commercially exploitable – national heritage form.

Can the same be said of nature writing itself? Jos Smith is among a number of critics to claim that recent nature writing in Britain has turned its back on what he calls those strands of 'preservationist nationalism' that used to thrive on nostalgic visions of Old England, and has cultivated instead a broad 'archipelagic' sensibility that is appropriately attentive to the interconnected histories of the British Isles.[167] Smith cites ample evidence in support, but the idea that British nature writing, especially *English* nature writing, has outlived its usefulness for the nation is at best wishful thinking and at worst an example of a critical tendency in the predominantly left-liberal Western academy to look right past the conservative ideologies that are sometimes embedded within even the most progressive-looking literary texts. A better argument to make – and the one made in this book – is that nature writing is neither a progressive nor a reactionary form per se, but one that can be mobilised for a variety of ideological interests; it is also a form that is singularly well suited to showing these conflicting interests at work.

This has obvious ramifications for the historical representation of weather in British nature writing, which by no means follows a logical progression that accords with the latest meteorological discoveries. Nor is it safe to say that, after the so-called ecological turn, weather is no longer seen as an external force but rather as an integrated exercise in the reciprocal if uneven exchange of material properties. The environmental philosopher Arnold Berleant describes this second process well when he claims that 'there is no inside and outside, human being and external world, even, in the final reckoning, self and other . . . Rather, we live in a dynamic nexus of interpenetrating forces to which we contribute and respond.'[168] More recently, the anthropologist Tim Ingold describes it even better when he refers to the cross-hatched phenomenological concept of the 'weather-world', a set of mutually transformative processes in which human beings both act upon and are acted upon by the various phenomenal 'worlds' through which they travel, and in which the different human and

[167] Smith, *New Nature Writing*, n.p.; for further 'archipelagic' perspectives, see Pippa Marland, *Ecocriticism and the Island: Readings from the British-Irish Archipelago* (Lanham, MD: Rowman & Littlefield, forthcoming 2022), and Elaine Stratton, Godfrey Baldacchino, Carol Farbotko and Andrew Howard, 'Envisioning the Archipelago', *Island Studies Journal*, 6 (2011), 113–30. For a further discussion of British island writing within an archipelagic context, see also Chapter 3.

[168] Arnold Berleant, *Aesthetics and Environment: Variations on a Theme* (Farnham: Ashgate, 2005), p. 13.

nonhuman actors intermingle, equally volatile forces that are neither earth-bound nor sky-bound but both of these in tandem, in a more or less constant state of flux for which 'atmosphere' seems the most appropriate word.[169] Such broadly new materialist understandings of 'distributed agency' (Bennett)[170] challenge some of the favourite strategies used to describe weather and its effects in English literature, notably the pathetic fallacy, which is now either seen as a relic of Romantic symbolic thought or, conversely, not a fallacy at all.[171]

Orthodox though these understandings have become for a new generation of eco-philosophers and critics, a gap remains between what ecocriticism says and what nature writing does – a gap that this book cannot hope to narrow, but that it tries at least to critically explore. To reiterate the argument of this chapter, 'lateness' names that gap, which has already appeared in this book in several different guises: a gap between experience and representation, a gap between language and knowledge, a gap between human consciousness and the so-called natural world. 'Lateness', in this sense, is not just – perhaps not even primarily – a temporal marker; rather it marks the uncertainty that is the lifeblood of modern British nature writing, which seeks in turn to inhabit the multiple contradictions that characterise the experience of living in the modern world. In what remains of this chapter, we glance at two contributions, by Melissa Harrison and Richard Mabey respectively, to the ongoing national conversation about 'English weather'.[172] As will be seen, this conversation is showing little sign of drying up even as the earth itself is warming. Indeed, part of its purpose is to sustain itself and its subject through 'the last years of English weather'[173] – through a late if not final period in which, locked as we are into a seemingly interminable present, we must all come to terms with permanent crisis without succumbing

[169] Timothy Ingold, *Being Alive: Essays on Movement, Knowledge and Description* (London: Routledge, 2011), pp. 96, 131–2; on the concept of 'atmosphere', see also Morton, *Ecology without Nature*, and, especially, Jesse Oak Taylor, *The Sky of Our Manufacture: The London Fog in British Fiction from Dickens to Woolf* (Charlottesville: University of Virginia Press, 2016).

[170] See Bennett, *Vibrant Matter*, p. 31.

[171] Neil Evernden, 'Beyond Ecology: Self, Place, and the Pathetic Fallacy', in Cheryll Glotfelty and Harold Fromm (eds.), *The Ecocriticism Reader: Landmarks in Literary Ecology* (Athens: University of Georgia Press, 1996), pp. 92–104. For further reflections on the use of pathetic fallacy in nature writing, especially within a Romantic context, see Chapter 1.

[172] Melissa Harrison, *Rain: Four Walks in English Weather* (London: Faber and Faber, 2016); Richard Mabey, *Turned Out Nice Again: On Living with the Weather* (London: Profile Books, 2013). Further references are in the text.

[173] Harris, *Weatherland*, p. 389.

to the false alternatives of technological optimism or existential despair.[174]

***

Melissa Harrison's short excursus *Rain* is organised round four country walks in different parts of England in different seasons, followed by two complementary glossaries of popular (folk-driven) and scientific (meteorological) words for rain. Its focus is on the mood-altering as well as landscape-altering properties of rain, which Harrison describes as both 'co-author of our living countryside' and 'part of our deep internal landscape . . . as essential to our sense of identity as it is to our soil'.[175] As Alexandra Harris argues for the English weather more generally, rain serves in Harrison's text as at once agent of material change and imaginative nutrient, with the 'pleasurable melancholy' it produces issuing a reminder that the age-old idea of 'weathering' functions as a mixture of stoical endurance and unexpected joy.[176] 'Weathering' happens to humans and nonhumans alike, suggesting that Harrison's text might be amenable to a new materialist reading in which weather acts as a bridge between different material entities, just as it acts to fold together things and words, the primary experience and the secondary representation of the natural world.[177] Certainly it's true that the text's four essays all inhabit regional 'weather-worlds' in which the elements are in flux, and a vivid impression is given of interconnected actors in quasi-perpetual motion, with 'wayfaring' – to cite Ingold's resonant term – being 'the fundamental mode by which living beings inhabit the earth'.[178]

However, this reading potentially supports rather than challenges Harrison's semi-mythic view of the 'English weather' returning us to ourselves as well as restoring our place within the natural world of which we are still an integral part.[179] This universal message contains or, perhaps better, conceals several others, suggesting the text's 'late' status as an early twenty-first-century work of English country writing that hides quietly nationalistic sentiments behind the stated desire to re-establish a 'fuller, older, and more deeply felt relationship to the natural world'.[180] Needless to say, this 'late' status is also conveyed in other ways, for example, through the text's intermittent references to climate change and its opening

174 See Donna Haraway, *Staying with the Trouble: Making Kin in the Chthulucene* (Durham, NC: Duke University Press, 2016).
175 Harrison, *Rain*, pp. xi–xii. 176 *Ibid.*, p. xii.
177 For other new materialist readings, this time of Romantic-period authors, see Chapter 1.
178 Ingold, *Being Alive*, p. 12. 179 Harrison, *Rain*, p. 65. 180 *Ibid.*, p. 88.

admission that 'we may all need to get used to more extreme weather conditions, more often, as the long hangover from the excesses of the industrial revolution begins to bite'.[181] Helen Macdonald is also right, in her review of the book, to point out that it looks to offer us 'a new kind of modern pastoral: peopled, raw, messy, shining' – though this is reminiscent less of Harrison's than of her contemporary Kathleen Jamie's work.[182]

A better comparison might be with John Lewis-Stempel, though Harrison, unlike Lewis-Stempel, is deliberately inclusive of her family and – like Jamie again, perhaps – is acutely conscious of the continuing division of labour that underpins bourgeois fantasies of 'the country life'.[183] As previously argued, *Meadowland* is conservative by *design*, even if it is aware of the changes daily wrought upon the English countryside; its dreamy attempts to register ancestral continuity are thus always balanced against its author's practical realisation that the farms and fields of England are working spaces that require the protection of property and profit as much as the stewardship of local wildlife. *Rain* is conservative by *default*, buoyed by its author's childhood memories and, still more, by translated memories of her late mother, who, born and brought up in India, always 'loved the English countryside – Dartmoor particularly – and was capable of a childlike rapture that embarrassed me as a teenager, but which shaped me nonetheless'.[184] That England, and English weather in particular, impresses itself upon both works should come as no surprise at a time when the idea of Englishness is increasingly challenged, but it is still disappointing to find contemporary works of British nature writing that manage paradoxically to feel so up to the minute and so out of date.

***

A similar mixture of stoicism and delight can be found in Richard Mabey's short book *Turned Out Nice Again: On Living with the Weather* (2013), which shares Harrison's view of weather as an integral part of British popular culture ('the one circumstance of life which we [have] in common' (7)), but a form of traditional folk knowledge that risks being lost ('we barely have a weather memory anymore').[185] For Mabey, as for Harrison, weather operates as a dense weave of metaphorical association and real experience, an amalgam of moods and memories that owes as much to accumulated human imagination as to immediate atmospheric conditions, but that also reminds us of our limited powers of interpretation – we may see in it what we want to see, but at times we may not see much at all, or feel instead 'what is missing from our lives' (p. 48).

[181] *Ibid.*, p. 4. [182] Helen Macdonald, cover blurb for *Rain*, n.p.
[183] See the discussion of Lewis-Stempel's *Meadowland* earlier in this chapter.
[184] Harrison, *Rain*, p. 72. [185] Mabey, *Turned Out Nice*, p. 52. Further references are in the text.

In this context, *Turned Out Nice Again* stages what Mabey calls 'our tragicomic street theatre of daily coping' (p. 90) with the weather: an accommodating process that involves creative adaptation to changeable local conditions, but also a longer-term recognition that our interpretation of what we are experiencing is often 'emotionally warped and highly selective', as are our memories of our own and other people's lives (p. 46). Mimicking these conditions, Mabey's work presents a kind of mini-compendium of assorted anecdotes and stories that look to match the vagaries of British weather, which itself supplies the 'meteorological equivalents of biodiversity, a tribute to the variety and eccentricity of Britain's landforms' (p. 28). This is British nature writing in one of its most readily identifiable forms, whimsical and contemplative by turns, and consciously operating within a national literary tradition that recasts the amateur naturalist as storyteller-collector and vice versa, with acknowledging nods to the previous 'weather writing' (p. 16) of the Romantics (Clare, Coleridge, the Wordsworths) and the Victorians (Jefferies, Ruskin), but, most of all, to the work of British nature writing's foundational figure, Gilbert White.

White, the subject of an award-winning biography by Mabey, is seen throughout the latter's work as an accomplished 'weather storyteller' who succeeded in capturing his own experiences of the natural world with an 'intimacy of detail' that was unrivalled in his time and has remained unmatched (pp. 14–15). Like White's, Mabey's writing seeks to restore a sense of wonder to everyday life, not least by registering human sensory perceptions to local atmospheric phenomena, or, perhaps better, by *intermingling* these perceptions and phenomena in such a way as to fashion 'weather-worlds' that are of our own making if not necessarily subject to our own control. But as also with White, there is a darker aspect to Mabey's writing that represents what he sees as the 'dark shadow behind [even] the most seductively dramatic and beautiful weather' (p. 36). One of the five interlinked chapters in *Turned Out Nice Again* is entitled 'Black Dog', and in it Mabey speaks candidly about his struggle with clinical depression – a struggle related at greater length in his best-known work, *Nature Cure* (2005). 'Melancholy' is probably the word that describes this condition best: a debilitating fluctuation between the equal-and-opposite moods of elation and despair.[186] Indeed, it is tempting to see all of Mabey's work as battling against its own melancholic tendencies. These tendencies, as we

[186] For a thoroughgoing discussion of melancholy that traces its historical roots as well as examining its contemporary conditions, see Jennifer Radden (ed.), *The Nature of Melancholy: From Aristotle to Kristeva* (Oxford: Oxford University Press, 2002).

have been suggesting in this chapter, are similarly apparent in a good deal of contemporary British nature writing, and their traditions go back, if not to White then to another foundational figure – and major influence on Mabey – the late Romantic peasant-poet, John Clare.[187]

Seen in this context, *Turned Out Nice Again* – like *Nature Cure*, perhaps – contains several levels of irony in its title, one of which pertains to the possibility that things will not only *not* turn out for the better, but may get considerably worse. This is nowhere better demonstrated than in the last chapter, 'The Storm Clouds of the Twenty-First Century', which is ostensibly a discursive engagement with the realities of contemporary global climate change, but which, much like the rest of the book, offers more of a personal reckoning with the author's dark Ruskinian visions of an irremediably blighted world (pp. 77–8). Mabey is careful to distance himself from Ruskin, as from (1) pessimistic visions of a warming world that is already well set on the path to its own inevitable destruction, and (2) technological fantasies of 'fixing' the problem.[188] Both of these he sees as contaminated by a popular media discourse that sensationalises as it simplifies, and that is often pitched between two alternative kinds of denialism – the kind that wilfully misunderstands the size and scale of the problem and the kind that imagines that humanity, having created it, can now solve it, with arrogant disregard for the intricate ways in which the earth actually works (pp. 82–3). However, it is clear that he shares at least some of Ruskin's fears, which he sees as 'chim[ing] with our current social anxieties' (p. 77). At the heart of these anxieties, Mabey says of Ruskin, 'was a frightened man ill-at-ease with himself, with the blind and insensitive advance of technology, with nature's seeming indifference to humankind, and our indifference to nature' (p. 77). Ruskin, he suggests, partly mirroring himself, was 'a man projecting onto the weather his feelings about the state of the world. Someone rather like ourselves, in fact, as we gaze at the gathering storm clouds of the twenty-first century' (p. 78).

Mabey's own response to the gathering gloom is to call for calm, accepting that climate change is real but also that it is not an unprecedented problem, and that 'we will probably have to confront [it] in the way the planet has always done, muddling and adapting our way through as best we can' (p. 83). This pragmatic view, politically quietist though it

[187] For a detailed discussion of Clare and his work, see Chapter 1.

[188] For an example of the former, see Roy Scranton, *Learning How to Die in the Anthropocene: Reflections on the End of a Civilization* (San Francisco: City Lights, 2015); for an example of the latter, see Robert Kunzig and Wallace S. Broeker, *Fixing Climate: What Past Climate Change Reveals about the Current Threat – And How to Counter It* (New York: Hill and Wang, 2009).

seems, has something to be said for it, even if the use of 'we' elides the unequal distribution of responsibility for climate change and susceptibility to its effects. It also accords well with recent interventions from Donna Haraway, whose latest book urges us to 'stay with the trouble', and the British geographer Nigel Clark, whose 2010 book *Inhuman Nature* offers a salutary reminder that the earth is not ours to fashion as we choose, and that 'alongside our capacity for action [exists] a kind of primordial passivity, a susceptibility in the face of all that is not ours to make or even know'.[189]

For both Haraway and Clark, the current global environmental crisis – of which global warming is just a part – calls for a response that goes beyond what we have been calling in this chapter Anthropocene thinking – that is, it requires concerted human intervention, but also a collective recognition that we are subject to elemental forces beyond the human capacity for management – that at one level are *nothing to do with us.* Another way of saying this is that, in their different ways, Haraway and Clark are suggesting that if there is no such thing as a 'return to nature', there is an imperative to engage with multiple *natures*, each of these enfolded within the others and all of them presided over by an autonomous earth. It remains an open question as to whether contemporary British nature writing has been able to confront these inherently volatile 'multinatures'.[190] To some extent, it still cleaves to a national literary tradition in which the immediate effects and consequences of weather have been systematically preferred to the speculative predictions of climate, though as this book has shown, the supposedly natural phenomena of weather and climate, along with the various personal and cultural associations mapped onto them, have been as wrapped up in one another throughout the modern history of British nature writing as they are in contemporary ecocritical accounts. To some extent as well, modern British nature writing has remained limited by its scope and by its imprisonment within the very sets of conventions and traditions from which it has periodically sought to free itself. This struggle is coeval with modernity itself, one irony of which is that it only appears to be new; another that it rebels against the conditions that it has created for itself.[191] And if modernity is a theatre of

189 Nigel Clark, *Inhuman Nature: Sociable Life on a Dynamic Planet* (London: Sage, 2010), p. 52; see also Donna J. Haraway, *Staying with the Trouble* (Durham, NC: Duke University Press, 2016).

190 See Lorimer, *Wildlife in the Anthropocene.*

191 Marshall Berman, *All That Is Solid Melts into Air: The Experience of Modernity* (New York: Simon and Schuster, 1982). See also Chapter 3.

conflict, so too is modern British nature writing, whose contemporary iterations are but the latest reminders that 'to be modern is to find oneself in an environment that promises growth, power, adventure . . . but that also threatens to destroy everything we have, everything we know, everything we are'.[192]

[192] *Ibid.*, p. 15.

Afterword
Shades of White

It seems appropriate in several respects to begin and end this book with Gilbert White, the humble but, as it would turn out, enormously influential parson-naturalist whose 1789 study *The Natural History of Selborne* has long since cemented its status as British nature writing's foundational work. There are multiple shades of White in much of the nature writing that followed him, and tributes from contemporary figures like Richard Mabey, Robert Macfarlane and Melissa Harrison abound. British nature writing in its traditional natural-historical vein owes a great deal to White, but a more nagging question is whether it is also a subsidiary form of 'white writing', indebted to ethnically circumscribed, if not necessarily acknowledged, ways of looking at the world.[1]

Is the history of British nature writing a pictorial exercise in shades of white? Some critics seem to think so. Certainly ecocriticism has often failed to address the problem. Rebecca Pohl, for instance, in her mostly generous review of Jos Smith's 2017 study *The New Nature Writing*, suggests that for all Smith's spirited attempts to make a case for the activist emphasis of contemporary nature writing, he falls into a familiar pattern in neglecting issues of gender, class and, especially, race.[2] British nature writing and ecocriticism are by no means unique in this regard, and similar things have been said in the United States where, as Carolyn Finney puts it, 'racism [has played] a significant part in limiting black participation in the "great outdoors"'.[3] But whereas steps have been taken in the United States to address this – the excellent critical work of Camille Dungy, Lauret

[1] J. M. Coetzee, *White Writing: On the Culture of Letters in South Africa* (New Haven, CT: Yale University Press, 1988).

[2] Rebecca Pohl, 'Book Review: *The New Nature Writing*: Rethinking the Literature of Place by Jos Smith', 11 November 2017, blogs.lse.ac.uk/serereviewofbooks/2017/11/09/book-review-the-new-nature-writing-rethinking-the-literature-of-place-by-jos-smith/ (accessed 9 October 2020).

[3] Catherine Buni, 'Toward a Wider View of "Nature Writing"', *Los Angeles Review of Books*, 10 January 2016, lareviewofbooks.org/article/toward-a-wider-view-of-nature-writing/ (accessed 10 October 2020).

Savoy and Finney herself comes to mind – British nature writing still has relatively few high-profile writers of any colour other than white. Similarly, despite powerful rallying calls from Raymond Williams (1973) and, latterly, Kathleen Jamie (2019), British ecocriticism until recently has been startlingly tone-deaf to class and gender differences. While Mark Cocker (2015) is surely wrong to accuse it and the creative writing it supports of being 'apolitical', it still tends either to be wary of *announcing* its politics or, even when it does, to fold social into environmental activism, with insufficient regard for the continuing race, class and gender divisions against which such activism seeks common ground.[4]

There are good historical reasons for this: the association of Britain's 'wild' places with displacement and exclusion; the reliance on literary modes and genres, especially the pastoral, that serve to naturalise social inequality; the unequal access to the time and space required to produce nature writing; the tendency (again until fairly recently) to concentrate on rural rather than urban environments, the latter of which are more likely to draw attention to the 'links between environmental disadvantage, race and class'.[5] None of this is anything like enough to justify Stephen Poole's swingeing claim that British nature writing is little more than a vehicle for 'bourgeois escapism', but the charge still leaves a sour taste, suggesting that popular perceptions of nature writing as a conservative reflex mechanism continue to have some traction regardless of what is actually written and read.[6]

As Richard Mabey, in a truculent rejoinder to Poole, insists, nature writing is nothing if not diverse, and even some of its most traditional features, on closer inspection, are far less traditional than they appear to be.[7] But the question remains as to whether it is diverse *enough*, raising vexed issues of inclusivity and representation, which exist in relation to the equally thorny problem of sociopolitical engagement. In the United

[4] Raymond Williams, *The Country and the City* (London: Chatto and Windus, 1973); Patrick Barkham, 'Interview. Kathleen Jamie: "New Nature Writing Has Been Colonised by White Men"', *The Guardian*, 17 October 2019, www.theguardian.com/books/2019/oct/17/kathleen-jamie-surfacing-interview-nature-writing/ (accessed 10 October 2020); Mark Cocker, 'Death of the Naturalist: Why Is the "New Nature Writing" so Tame?', *New Statesman*, 6 June 2015, newstatesman.com/culture/2015/06/death-naturalist-why-new-nature-writing-so/tame/ (accessed 10 October 2020).

[5] Kate Oakley, Jonathan Ward and Ian Christie, 'Engaging the Imagination: "New Nature Writing", Collective Politics and the Environmental Crisis', *Environmental Values*, 27 (2017), 687–705 (p. 698).

[6] Stephen Poole, 'Is Our Love of Nature Writing Bourgeois Escapism?', *The Guardian*, 6 July 2013, www.theguardian.com/books/2013/jul/06/nature-writing-revival/ (accessed 10 October 2020).

[7] Richard Mabey, 'In Defence of Nature Writing', *The Guardian*, 18 July 2013, www.theguardian.com/.../18/richard-mabey-defence-nature-writing/ (accessed 10 October 2020).

Kingdom as in the United States, there is a continuing racial bias in environmental organisations that suggests that not enough effort is being made to enlist black and ethnic minority Britons for environmentalist causes, at heart a *social* problem that reflects a residually divided society, albeit one that is less conspicuously stratified in the United Kingdom than it is in the United States. To some extent, the problem is also one that revolves around what Charles Taylor calls the politics of recognition.[8] Including 'non-white' Britons is one thing, *recognising* their achievements is another, while still another is recognising the many occasions on which they have been marginalised or left out.

But this isn't to say that British nature writing and ecocriticism are or have been exclusively 'white', and the contemporary period alone features several established or increasingly well-known black figures, among them the writer Elizabeth-Jane Burnett (see later in this afterword), the poet Benjamin Zephaniah, the photographer Ingrid Pollard and the television presenters Patrick Aryee and Liz Bonnin (the democratisation of 'TV nature', while still a work in progress, is more evident than its literary counterpart, at least in the United Kingdom). At the same time, a clutch of younger British Asian writers, Jini Reddy for instance, is increasingly coming to prominence, while even some grizzled veterans like the late Trinidad-born author V. S. Naipaul can stake some claim to having engaged with environmental issues and debates.[9] Still, it would probably be fair to say that British nature writing today remains a predominantly 'white' form, albeit one increasingly minded to interrogate its own biases. More contentiously perhaps, it is also *regionally* dominated, with many of the best-known writers living in East Anglia, and several of them, friendly with one another, supporting each other's work. It would no doubt be going too far to describe this as a club, but there is a mutual affability to some of the exchanges that can be irritating to those who happen to come from other regions of England, or Wales, or Scotland, or other parts of the British Isles. Some of these writers double as critics, which merely intensifies the regional effect, while the vast majority are men, lending strength to the view, expressed most forcefully by Astrid Bracke, that British ecocriticism has yet to give women writers, with some notable exceptions, the recognition they deserve.[10]

[8] Charles Taylor, *Multiculturalism and the Politics of Recognition* (Princeton, NJ: Princeton University Press, 1992).

[9] See, for example, V. S. Naipaul, *The Enigma of Arrival* (New York: Viking, 1987).

[10] Astrid Bracke, 'Macho Nature? Or, Gender in New Nature Writing Part I', 19 February 2013, astridbracke.com/2103/02/19/macho-nature-or-gender-in-new-nature-writing-part-i/ (accessed 10 October 2020); also Bracke, 'Typically Feminine? Or, Gender in New Nature Writing

This lack of recognition seems to have affected one of the most intriguing recent works of British nature writing: Elizabeth-Jane Burnett's *The Grassling: A Geological Memoir* (2019). To the best of our knowledge, it received only one broadsheet review and was not even longlisted for the Wainwright Prize for UK Nature Writing.[11] In such a crowded field, of course, it is often difficult for new voices to be heard. But Burnett, of English and Kenyan heritage, was already an established poet when she wrote *The Grassling*, which was published not by an obscure independent press, but by Penguin's prestigious hardback imprint Allen Lane. Whether the continuing dominance of white masculinity in British nature writing has been a factor in *The Grassling*'s relative neglect is impossible to know. That may well be the case, but another factor may be the text itself. Burnett's memoir of her father's decline and death, and the long history of his family in the village of Ide in Devon, is a much *stranger* book than most of its competitors. This is despite the familiarity of its concerns: rootedness, locality, memory, natural history, self-realisation and self-forgetting. It particularly explores the long-standing and deep-seated Burnett connection to the land. Such connections can seem troubling for the exclusions they imply, and yet the book finds them creatively liberating. There are a few scattered references to Burnett's Kenyan roots, including a telling passage in which she describes how, despite the fact that through her father she 'stretches back generations' in Ide, 'while others see him as belonging, even without knowing his story, they do not see that in me'.[12] But what really sets the text apart is its obliqueness and formal complexity; it is modern*ist* to an extent that is unusual in British nature writing (see Chapters 3 and 4). Some chapters offer relatively conventional nature memoir, but they are interspersed by more impressionistic, poetic sections ('EIGHTY YELLOWS for my father's eightieth birthday', 'SOIL MEMOIR FOR DRUID'S HILL'). The book's second part also contains chapters in which the narrator's voice merges with that of the mysterious Grassling: a plant entity which both is and is not a figure for Burnett herself; which is utterly immersed in local ecologies, but also constituted by words; which is deep-rooted in the soil, but reaches up to the stars.

Part 2', 1 January 2014, astridbracke.com/2014/01/01/typically-feminine-gender-in-new-nature-writing-part-2/ (accessed 10 October 2020).

[11] It received a short, positive review by P. D. Smith in *The Guardian* (21 March 2019); it was also reviewed in the *Times Literary Supplement* (7 June 2019). As of 4 December 2020, the paperback stands at number 210,614 in the Amazon 'Best Sellers Rank', compared (for example) to number 1,624 for the paperback of another 2019 book, Robert Macfarlane's *Underland*. We are grateful to Kat Waters for a helpful discussion of *The Grassling*.

[12] Elizabeth-Jane Burnett, *The Grassling: A Geological Memoir* (London: Penguin, 2020), p. 47.

The book's complex, non-linear structure is matched by its linguistic playfulness. Even its most workaday passages are stylistically rich, epigraphs take the form of abstract poetry, and long sections of prose poetry attempt to evoke sensory immersion in the nonhuman world through an array of formal devices: truncated sentences, parataxis, figuration, repetition, typographical variation and onomatopoeia. Like earlier nature writers such as John Clare, Burnett is fascinated by the capacity of language to evoke the nonhuman world through mimicry: 'For a while it is a blackbird. Trill-la-la-lickwrrrrrrrrrrrrrrrrr rrrrrrrrrrrrrrrrrr.'[13] The text's fluidity is also apparent in shifts in narratorial perspective that complicate its status as a first-person memoir: indeed, at times Burnett's writing is reminiscent of the Virginia Woolf of *Mrs Dalloway* (1925) or *The Waves* (1931). The result is rich, powerful and disorientating. If at times the book seems overwrought, its willingness to take formal risks is certainly to be applauded, especially given British nature writing's general tendency, as noted in previous chapters, to play it safe.

A recent risk-taking book with which it has a surprising amount in common is Charles Foster's *Being a Beast* (2016: see Chapter 4); indeed, Burnett's might have been subtitled *Being a Plant*. Although her book is more stylistically innovative, like Foster, she strains imagination, language and, at times, credulity in her pursuit of a non-anthropocentric perspective. Whereas Foster's self-ironising reflections on the limitations of his approach are very much to the fore, Burnett's are buried deeper. Nonetheless, the book is partly about its own composition and self-consciously addresses the intimacies and the estrangements of ecomimesis. In its moving final chapter, which addresses her father's death, Burnett's meditation on the lettings go that life requires of us seems to apply to nature writing itself: 'When you realise there is no possession, there is only drawing near.'[14] The book ends with the narrator writing 'slowly, the names of my family on the grass', a difficult activity as the 'ink won't settle' and her 'fingers numb and lock', before she merges into the landscape ('I am all dew all field all hope').[15] Taken on its own, the passage might be seen as a rather too neat reworking of the old ecomimetic fantasy (see Chapter 1). However, in the context of the book as a whole, it reveals nature writing as a joyous, agonising process of becoming other, which draws its power from the fact that it never quite reaches its destination.

Burnett's work, even if not yet fully appreciated, may indicate that the landscape of published British nature writing is slowly changing.

[13] *Ibid.*, p. 139. [14] *Ibid.*, p. 186. [15] *Ibid.*, p. 187.

At a grassroots level, *The Willowherb Review*, founded in 2019 by Jessica J. Lee (a British and Canadian writer of Taiwanese and Welsh heritage), provides an online platform for both emerging and established nature writers of colour.[16] Lee's reflections on the sheer volume and quality of contributions confirm that it is not the case, as has often been assumed, that there are few non-white nature writers. Rather, it is that their voices have rarely been heard: 'No one need ever ask where the nature writers of colour are; they're here in their hundreds.'[17] The Nan Shepherd Prize was launched in 2019 by publishers Canongate, in honour of Shepherd's inspirational legacy (see Chapter 3), with the aim of raising the profile of nature writers from under-represented communities. Meanwhile 2020, the conflict-ridden year that finally brings this book to a close, saw the high-profile signing of Anita Sethi to Bloomsbury for a trilogy beginning with *I Belong Here: A Journey along the Backbone of Britain.* Sethi describes the work as an 'act of resistance', at least partly in response to her experience as the victim of a race hate crime in 2019.[18]

It is not the case, however, that the doors to a more diverse genre are simply opening up, especially in relation to rural writing. As we have seen, Burnett writes of her sense that being deemed to 'belong' in the British countryside is still largely dependent on skin colour. In *Wanderland*, Jini Reddy, in addition to expressing her feeling that her explicitly spiritual quest to connect with the natural world is at odds with the secularism of contemporary nature writing, alludes to being the subject of 'that look' when venturing beyond cosmopolitan London.[19] She reflects on her uncomfortable awareness that 'people in the countryside or even suburban parks don't often come across a woman with brown skin walking on her own', but, even so, she acknowledges her own relative sense of privilege: 'Suddenly I wonder what it must feel like to be a black man who enjoys going for hikes in the woods, any woods. It's that feeling of being conspicuous, of being othered.'[20] Writers like Burnett, Reddy and Sethi are

16 www.thewillowherbreview.com (accessed 23 January 2021).

17 Jessica J. Lee, *The Willowherb Review* Twitter account, 11 June 2020.

18 www.thebookseller.com/news/bloomsbury-signs-sethi-powerful-new-voice-nature-writing-1206731 (accessed 23 January 2021). Sethi words chime with the experiences dramatised in the play *Black Men Walking*, written by rapper and beatboxer Testament, in which a group of black men walking in the Peak District reflect on the 'suspiciousness' with which they are viewed in the rural environment (London: Oberon Books, 2018), p. 25. The play also offers a corrective to narratives of a uniquely white rural history by referencing the presence in the landscape of black figures from the past.

19 Jini Reddy, *Wanderland: A Search for Magic in the Landscape* (London: Bloomsbury, 2020), p. 29.

20 *Ibid.*, pp. 59–60.

part of the likely future of nature writing in the United Kingdom, but social obstacles remain, and they have still needed the courage to stake their claim to belonging, both in the rural landscapes they inhabit and in the cultural milieu of nature writing itself.

A similar observation might be made of contemporary LGBTQ^{+} nature writers, who are becoming increasingly visible in the genre. Recent works such as Luke Turner's Epping Forest–based meditation on masculinity and bisexuality, *Out of the Woods* (2019), and Alys Fowler's *Hidden Nature* (2017), which offers parallel accounts of her kayaking the canals of Birmingham and coming out as a gay woman, do much to challenge the heteronormativity of British nature writing. Meanwhile, Mike Parker's *On the Red Hill: Where Four Lives Fell into Place* (2019) opens gloriously with an account of the first same-sex civil partnership ceremony performed in the Powys Registry Office: an event which formalises the relationship of Reg and George, two men who have been living together for nearly sixty years. The episode functions as a kind of performative inauguration of contemporary gay rural writing. But Parker's book does more than carve out a new space in the genre for sexual diversity; it also challenges the prevailing assumption Parker himself identifies in gay culture, that the countryside is simply a place from which to flee. 'If the countryside appears at all in gay histories', he writes, 'it is usually only as a place to escape from, and as swiftly as possible'.[21] On the contrary, Parker's multigenerational story of the lives of four gay men in a Welsh village acknowledges the existence of a long-standing 'queer rural' history.[22] The cultural touchstone for this, as the epigraph to the book indicates, is E. M. Forster's early twentieth-century story of homosexual love, *Maurice*, with its eponymous character's conviction that 'Men of my sort could take to the greenwood.'[23]

Parker expresses his own love of the countryside as a deep-seated desire, one so powerful that it 'would have been dangerous to resist, like the act of coming out itself'.[24] His decision to move to rural Wales was informed by his sense that, contrary to his 'gobby gay Brummie' acquaintance's prediction that he would either be ignored or actively persecuted, there was a 'greenwood' of sorts to be found there: 'On my travels in Wales, even out to its wildest corners, I'd often sniffed a queer cunning in the air. A quiet tolerance too.'[25] Where Burnett's book is linguistically innovative, *On the Red Hill* is distinguished by its formal hybridity. It combines passages of biography, memoir and topographical celebration, interspersed with poems and diary entries,

[21] Mike Parker, *On the Red Hill: Where Four Lives Fell into Place* (London: Windmill Books, 2019), p. 5.
[22] *Ibid.*, p. 6. [23] *Ibid.*, p. 2. [24] *Ibid.*, p. 6. [25] *Ibid.*, p. 7.

paintings and drawings, historical documents and black-and-white photographs, some recent, others dating back to the older men's youth. Nevertheless, the natural world is at the heart of the book, and its sections are arranged according to the elements, the seasons and the four cardinal points of the compass, as well as offering biographies of the four men. The cumulative effect is one in which nature and feeling for nature are inseparable from ideas of human love and human community.

Attesting to how rapidly things can shift given sufficient momentum, the 2020 Wainwright Prize shortlist was notable for its diversity. Parker's *On the Red Hill* and Reddy's *Wanderland* featured on the list, as did other works that signal new directions in British nature writing. David Gange's *The Frayed Atlantic Edge* (2019) speaks to the developing archipelagic mode, in which the view from the margins is pre-eminent (see also Chapter 3).[26] Gange's deliberately slow, inductive method of travelling the Atlantic seaboard of Britain and Ireland by kayak and on foot uncovers narratives that challenge mainland, urban-centric histories. Patrick Laurie's *Native* (2020) reflects the 'new georgic' of wilder farming, which combines traditional methods with an informed understanding of environmental risk and benefit, and contributes to the valuable persistence in nature writing of the voice of those who *work* the land as well as those who *walk* it.[27] The 2020 Wainwright Prize was eventually won by Dara McAnulty, at fifteen years old the youngest ever recipient of the award, with his *Diary of a Young Naturalist* (2020). This memoir, which takes the reader through a calendar year of the author's life in nature, is significant in a number of ways. McAnulty is autistic, and the book is in part an eloquent means of conveying his distinctive perspective on the natural world, as well as some of the personal difficulties he encounters along the way. But it also articulates a cry of frustration that might be seen as representative of the youth environmental movement more generally. In the current environmental crisis, and in the context of the 'power struggle between huge organisations, economics, development and the species we share our planet with', McAnulty writes, 'it's easy to become overwhelmed, depressed and disconnected'.[28] This sense of disconnection was

[26] David Gange, *The Frayed Atlantic Edge: A Historian's Journey from Shetland to the Channel* (London: William Collins, 2019).

[27] Patrick Laurie, *Native: Life in a Vanishing Landscape* (Edinburgh: Birlinn, 2020). For discussion of the 'new georgic', see Pippa Marland, 'Rewilding, Wilding, and the New Georgic', *Green Letters: Studies in Ecocriticism*, 24 (2020), 421–36.

[28] Dara McAnulty, *Diary of a Young Naturalist* (Ford: Little Toller, 2020), p. 199. Another popular work inflected by the author's autism is Chris Packham's *Fingers in the Sparkle Jar: A Memoir*

exacerbated in a pandemic year in which many, in the United Kingdom and elsewhere, suffered high anxiety and worse as a result of extended confinement, but which paradoxically saw a revitalised interest in the natural world.

With this interest in mind, Lee sees the 2020 Wainwright shortlist as evidence of the beginning of a new 'golden age' of nature writing, and is 'excited to be finally witnessing a shift in who we imagine a nature writer might be, and who we think may be reading their work'.[29] But while nature writing may be diversifying, access to much of Britain's nature remains in the hands of a privileged few. Nick Hayes' timely *Book of Trespass* bemoans the lack of a right to roam: 'By law of trespass, we are excluded from 92 per cent of the land and 97 per cent of its waterways, blocked by walls whose legitimacy is rarely questioned.'[30] The Covid-19 crisis, which at the time of writing still dominates our lives, has brought the problems of such unequal access to the fore. At the point when, as a nation, we realised more than ever before how much we valued and yearned for contact with nature, we were made aware of the pressure this placed on the limited number of places to which we had legitimate access; we were made aware too of the limited number of us who often like to speak on behalf of the nation as a whole. As Isabella Tree has noted in a recent article in *The Guardian*, the effects of the 'lockdown surge', once restrictions were lifted a little during the summer of 2020, would put unprecedented strain on the United Kingdom's national parks and nature reserves, threatening their vital role in providing habitat for what is left of the country's dwindling biodiversity.[31] Tree also uses the article to express her reservations about the 2020 campaign for extending the right to roam, led by the likes of Hayes, while acknowledging this as a 'crucial aim'.[32] There is a certain irony in her argument, notwithstanding the extraordinary success of the

(London: Ebury Press, 2017), which was adjudged to be the nation's favourite nature book in a 2018 poll organised by the UK Arts and Humanities Research Council (AHRC) in conjunction with the University of Leeds–based 'Land Lines' research team. For further information about the poll, see https://ahrc.ukri.org/favouritenaturebooks (accessed 23 September 2021); see also the 'Land Lines' project website at https://landlinesproject.wordpress.com/ (accessed 23 September 2021).

29 inews.co.uk/culture/books/wainwright-prize-nature-writing-winners-2020-new-diverse-voices-autism-640478 (accessed 23 January 2021).

30 Nick Hayes, *The Book of Trespass: Crossing the Lines That Divide Us* (London: Bloomsbury, 2020), front matter. Hayes discusses the different situation in Scotland, where the 2003 Scottish Land Reform Act extended public access rights (p. 368).

31 Isabella Tree, 'Lockdown Awakened Our Interest in Nature, but It Mustn't Be at the Expense of Wildlife', *The Guardian*, 28 December 2020, www.theguardian.com/commentisfree/2020/dec/28/lockdown-nature-expense-wildlife (accessed 1 February 2021).

32 *Ibid.*

Knepp wilding project (the subject of her 2018 book *Wilding*), given that she and her husband, 10th Baronet Charlie Burrell, live on an estate of 3,500 acres that has been in the Burrell family for generations. Nevertheless, her point that the depleted nature of the United Kingdom may not be robust enough to accommodate the population's appetite for it is an important one, indicative of the need for major systemic change. This includes, as she suggests, 'upgrading our 140,000 miles of footpaths [in England and Wales] and connecting them with urban areas' while also creating or restoring further extensive areas of wildlife-rich habitat.[33]

What general conclusions can we draw in 2022, at a worryingly isolationist time when Britain, which has cast itself adrift from the European Union, currently has one of the worst records of Covid-induced fatality in the world? One obvious judgement is that the United Kingdom is more socially riven than ever, with implications for both the natural environment and nature writing. If access to the land remains problematic and unequal, then so too is access to nature writing. Class is key here. Individuals with wealth, property, leisure, contacts, cultural capital and control over their schedules have a distinct advantage in all forms of cultural production, but particularly so when it comes to nature writing. This is in part because of the time and space required for careful observation of the natural world, as well as the energy needed to generate the intimate sense of encounter with the nonhuman that is at nature writing's core. Of course, working-class writers do come through – and we have discussed some key figures in this book – but it is very much against the odds.[34] For that reason, the recent shift from seeing nature as a wild place 'out there' to addressing its entanglements with our daily lives (including in urban spaces or edgelands) is very welcome. As we have emphasised throughout this book, the history of modern British nature writing is also a history of inequality. These intertwined histories run in tandem with enclosure, colonialism and the construction of knowledge and appreciation of 'nature' as a form of middle-class cultural capital. For Victorian nature writers – Ruskin and Miller, Hudson and Kingsley – nature studies represented a means of self-cultivation and an antidote to the philistinism of burgeoning urban mass culture (see Chapter 2). The depopulation of the countryside via enclosure and industrialisation allowed for its secrets and

[33] *Ibid.*

[34] There is increasing awareness of this issue; see, for example, the institution in 2020 of a prize for working-class nature writing, devised by Natasha Carthew: www.thebookseller.com/news/natasha-carthew-launches-working-class-nature-prize-1201897 (accessed 6 February 2021).

pleasures to be rediscovered by the bourgeoisie and transformed into elements of social distinction (despite the rhetoric of inclusiveness within which these discoveries were framed). Going back further, we should recall that Gilbert White was himself born into a relatively wealthy family, educated at Oxford, and able to practise natural history because of the generous free time his clerical role allowed. Since then, British nature writers have often presented themselves as amateur connoisseurs of the natural world, uniquely positioned to discover shades of beauty and significance beyond the knowledge of the field labourer, tourist or professional scientist. Nature writers, as we have seen in this book, offer far more than facts; they are also capable of offering powerful visions. As Ruskin noted, the value of such visionary insight derives from its rarity; it is supposedly possessed by a privileged, sensitive few. In this sense – and irrespective of individual authors' politics – elitist elements are baked into anything we might wish to identify as a tradition of British nature writing. These elements need to be contended with rather than simply ignored.

Index

CPSIA information can be obtained
at www.ICGtesting.com
Printed in the USA
LVHW020221260422
717218LV00006B/263

9 781107 191327